Mapping the Renaissance World

MAPPING THE RENAISSANCE WORLD

*The Geographical Imagination
in the Age of Discovery*

FRANK LESTRINGANT

Translated by David Fausett

With a Foreword by Stephen Greenblatt

Polity Press

First published in France as *L'atelier du cosmographe* © Éditions Albin Michel S.A., 1991

First published in 1994 by Polity Press in association with Blackwell Publishers. This book was published with the assistance of the French Ministry of Culture.

Editorial office:
Polity Press
65 Bridge Street
Cambridge CB2 1UR, UK

Marketing and production:
Blackwell Publishers
108 Cowley Road
Oxford OX4 1JF, UK

ISBN 0 7456 1147 8

A CIP catalogue record for this book is available from the British Library

Typeset in 10½ on 12 pt Garamond Stempel
by Graphicraft Typesetters Ltd., Hong Kong
Printed in Great Britain by T. J. Press Ltd., Padstow, Cornwall

This book is printed on acid-free paper.

Contents

List of Illustrations   vi
Foreword by Stephen Greenblatt: Thevet's Island   vii
Preface to the English language edition   xvi

Overture: Renaissance and Cosmography   1
1   The Cosmographical Model   12
2   Ancient Lessons: A Bookish Orient   37
3   Mythologics: The Invention of Brazil   53
4   Mythologics II: Amazons and Monarchs   71
5   Cartographics: An Experience of the World and
        an Experiment on the World   104
Epilogue: The End of Cosmography   126

Appendix: Extracts from Guillaume Le Testu's
        Cosmographie Universelle   132
Notes   136
Bibliography of Works by André Thevet   180
Index   187

Illustrations

Between pp. 110 and 111

1   Thomas de Leu, Portrait of André Thevet
2   Plate from Peter Apian, *La Cosmographie*
3   Hans Stradan, *The Triumph of Magellan*
4   Étienne Delaune, *Mêlée of Naked Warriors*
5   Antoine Jacquard, *The Cannibal*
6   *Savages in Combat*, from André Thevet's *Cosmographie Universelle*
7   *How the Amazons Treat Prisoners*, from André Thevet's *Singularités*
8   *The Ruse of Quoniambech*, from André Thevet's *Cosmographie Universelle*
9   *New Found Lands, or Isles of* Molues, from André Thevet's *Grand Insulaire*
10 *The Isles of Sanson, or of Giants*, from André Thevet's *Grand Insulaire*

Foreword

Some years ago a review of a book about the Tupinamba Indians in what is now the Bay of Rio in Brazil remarked that while virtually the entire population had been wiped out within a few generations after contact with Europeans, they had been 'extremely fortunate in their early ethnographers'. We should all be spared such good fortune. In the case of the Tupinamba, it principally took the form of two extraordinary French observers, Jean de Léry and André Thevet. Léry's account of his few months among the cannibals, *Histoire d'un voyage fait en la terre du Brésil* ('History of a Voyage to Brazil'), was first published in 1578 and has become an anthropological classic: in our century Claude Lévi-Strauss has paid homage to it as the 'breviary of the anthropologist'.[1] Thevet's main account of his even shorter stay, *Les Singularitéz de la France antarctique* (first published in 1557 and translated into English by Thomas Hacket in 1568 as *The Newfound Worlde, or Antarctike*) has in its own way also become a classic, but of a very different kind. Initially celebrated by the poets of the Pléiade as the French Jason who had brought the golden fleece of a whole new world back to his king and country, Thevet came under increasing attack in his own lifetime as a plagiarist, an impudent fool and a liar.

Such charges were scarcely disinterested. They emerged from the bitter, indeed murderous, environment of sectarian rivalry and hatred that tore France apart in the latter half of the sixteenth century (and, not coincidentally, doomed the French colony in Brazil). Léry, a Protestant pastor, attacks Thevet, a Franciscan monk, in terms that explicitly reflect the preoccupations of the religious wars, and Thevet's attacks on Léry are similarly motivated. But Thevet, who became *aumonier* to Catherine de Medici and Royal Cosmographer to the Valois kings, had the distinction of being attacked by fellow Catholics as well as Protestants. In the quarrels that swirled around his work there are the signs not only of doctrinal struggles but also of momentous shifts in the European understanding of what it meant to chart, describe and analyse the physical world and its human inhabitants.

That Thevet rather than Léry is the central figure of Frank Lestringant's *Mapping the Renaissance World* tells us a great deal about the project of this book. 'The Geographical Imagination in the Age of Discovery', Lestringant's subtitle and the object of his investigation, is less the origin of modern cartographic and ethnographic practices than it is the key to a vanished world, a world with a different set of interests and anxieties, a different standard of proof and disproof, a different sense of scale. That world, with its distinctive preoccupations and epistemology, began to change during Thevet's own lifetime, so that his works, once so central, began to drift uneasily towards an increasingly eccentric isolation. By the next century they had fallen into almost complete neglect, and the great Thevet was remembered, if he was remembered at all, as a fraud.

But *Mapping the Renaissance World* is not simply important as a piece of historical reconstruction. In the past few years Thevet has enjoyed something of a revival, a renewed interest whose sources are only partly antiquarian. Sceptical critiques of the conventions that underlie our own maps of the world have made it possible to call into question the positivist teleology that ranked all previous efforts by their relative proximity to or distance from strict scientific accuracy. Similarly, a growing awareness of the rhetorical conventions and complex purposes that shape anthropological writing in our own age has enabled us to investigate with something other than derision the conventions and purposes that governed in other ages the description of places and cultures. These changes in our intellectual climate are exquisitely articulated by Italo Calvino's postmodern fiction *Invisible Cities*, with its playful, disconnected descriptions of cities that jumble the features of the known and the unknown, systematically inverting the familiar and domesticating the strange, mingling memory and desire, fear and longing.

Calvino's Marco Polo entertains the Great Khan with accounts of cities that he has actually visited, cities that he has only read about and cities that he has invented – and, as the tales unfold, the distinctions seem increasingly untenable. The Great Khan, Calvino tells us, 'owns an atlas whose drawings depict the terrestrial globe all at once and continent by continent, the borders of the most distant realms, the ships' routes, the coastlines, the maps of the most illustrious metropolises and of the most opulent ports'. He leafs through the pages of his atlas in order to put Marco Polo to the test. The distinguished traveller recognizes some of the cities from first-hand experience and clear signs – Constantinople, Jerusalem, Samarkand; for others 'he falls back on descriptions handed down by word of mouth, or he guesses on the basis of scant indications.' The atlas contains still other cities of whose existence neither Marco Polo nor any other geographer is aware, 'though they cannot be missing among the forms of possible cities'. But even here the traveller is not at a loss: 'For these, too, Marco says a name, no matter which, and suggests a route to reach them. It is known that names of places change as many times as there are foreign languages; and that every place can be reached from other places, by the most various roads and routes.'² We are, without quite knowing it, re-entering the world of André Thevet.

But Calvino's work is a fiction, self-consciously and deliberately distanced from any known world. Thevet and the other mid-sixteenth-century geographers discussed in *Mapping the Renaissance World* insist, with an earnestness that often topples into belligerence, that they are representing reality itself, a reality that they claim to have seen in most cases with their own eyes. In 1595 Sir Walter Ralegh took a copy of Thevet's *Singularitéz* with him on his voyage to Guiana and verified what must have appeared to European readers one of the most improbable of traveller's tales: oysters that grow on trees. 'One salt river', Ralegh writes, 'had store of oisters upon the branches of the trees, and were very salt and well tasted.' 'This tree is described by Andrew Thevet', he adds, 'in his French Antarctique, and the forme figured in the booke as a plant very strange, and by Plinie in his 12. booke of his naturall historie. But in this yland, as also in Guiana there are very many of them.'³ What could have seemed like a palpable sign of lying – Thevet's recycling in a description of the New World of a marvel reported in an ancient text notoriously full of wild improbabilities – is certified as the plain truth by Ralegh's own eye-witness observation.

Even when such direct proof eludes him, Ralegh finds himself uneasily confirming or at least seriously entertaining the still more improbable claims of the fourteenth-century traveller Sir John Mandeville. Mandeville had reported the existence of a race of people whose heads grew beneath their shoulders: 'there are ugly folk without heads, who have eyes in each shoulder; their mouths are round, like a horseshoe, in the middle of their chest.'⁴ Cosmographers conventionally located such creatures – along with *Sciapods* (creatures with one enormous foot), Pygmies, hairy wild men, Amazons, and the like – at the extreme edges of the world, and Ralegh hears of exactly such people in Guiana: 'they are called Ewaipanoma: they are reported to have their eyes in their shoulders, and their mouthes in the middle of their breasts, and that a long traine of haire groweth backward betweene their shoulders'.⁵ To be sure, in this case, Ralegh cannot personally attest to their existence. He only chanced to hear of them, he writes, after he left the region; 'if I had but spoken one worde of it while I was there, I might have brought one of them with mee to put the matter out of doubt.' Ralegh is glancing here at the familiar European practice of bringing samples, including human samples, back from the newly discovered lands, in order to allow those at home to savour the wonders and confirm for themselves the truth of the travellers' tales. If eye-witnessing has been frustrated on this occasion, Ralegh is still inclined to believe in the existence of the Ewaipanoma: 'Such a nation', he writes,

> was written of by Mandevile, whose reports were holden for fables many yeeres, and yet since the East Indies were discovered, we find his relations true of such things as heretofore were held incredible: whether it be true or no, the matter is not great, neither can there bee any profit in the imagination; for mine owne part I saw them not, but I am resolved that so many people did not all combine, or forethinke to make the report.[6]

The queasy confusions of this long last sentence, lurching from scepticism to wonder to professed indifference and then to resolved acceptance, will repay some attention as a compressed index of the geographical imagination explored with so much subtlety and learning in Lestringant's book. Ralegh is careful to indicate that he knows that *Mandeville's Travels* have lost their credibility and that what were once taken to be accurate reports of distant lands have long been dismissed by informed readers as 'fables'. But the vast expansion of European knowledge of the rest of the world (exemplified for an Elizabethan Englishman by the circumnavigations of the globe by Drake and Cavendish) has confounded the secure boundary between fable and reality, verified what seemed incredible, and renewed the old cosmographic dreams.

It is in just this spirit that Ralegh's friend, the poet Edmund Spenser, responds to charges that the tales in *The Faerie Queene* are 'painted forgery, / Rather than matter of just memory'. Spenser urges his readers to reflect on the discoveries of great regions that had until recently been unknown to exist:

> Who ever heard of th'Indian *Peru*?
> Or who in venturous vessel measured
> The *Amazons* huge river now found trew?
> Or fruitfullest *Virginia* who did ever vew?
>
> Yet all these were, when no man did them know:
> Yet have from wisest ages hidden beene:
> And later times things more unknowne shall show. (*Faerie Queene* II.Proem. 2–3)

The blunt ethnocentrism of these lines – 'when no man did them know' – would probably not have disturbed Ralegh, but he might well have reflected uneasily on the fact that by this logic anything, however blatantly fabulous, could be justified and said to exist. (Spenser goes on to invoke the possibility of 'other worlds' on the moon or stars.) Ralegh had no desire to present his report on Guiana as a poet's dream, and he hastens to distance himself from the rehabilitation of Mandeville on which he had briefly embarked: 'whether it be true or no, the matter is not great, neither can there bee any profit in the imagination.' There might, the words may imply, be some profit in the reality, for people at home would certainly pay to see a man whose head grew beneath his shoulders. 'Were I in England now', says a traveller upon seeing the savage Caliban in Shakespeare's *The Tempest*, 'not a holiday fool there but would give a piece of silver: there would this monster make a man; any strange beast there makes a man: when they will not give a doit to relieve a lame beggar, they will lay out ten to see a dead Indian' (2.2.28–34). But (except for a gifted playwright) there is no profit in a mere story.

Ralegh assures his readers then that he knows the difference between reality and imagination (an important but difficult distinction for anyone in search of El Dorado), just as he knows the difference between great matters and small. And yet he does not let the subject rest there: 'for mine owne part I saw them not, but I am resolved that so many people did not all combine, or forethinke to make the report.' In the wake of the mingled scepticism and belief, the cross-currents of empiricism and imagination, we glimpse one of the key principles of the Renaissance geographical imagination: eye-witness testimony, for all its vaunted importance, sits as a very small edifice on top of an enormous mountain of hearsay, rumour, convention and endlessly recycled fable.

Hence it is altogether appropriate that Ralegh carries Mandeville with him to South America, for though the Elizabethan traveller inhabits a far different world from that of his medieval predecessor and is one of the avatars of English colonial exploration and settlement, he, like Thevet, shares much with the older cosmographical practice. When Ralegh or Thevet insist on what they themselves have seen with their own eyes, they are not in fact distancing themselves from this older practice so much as reproducing its traditional and time-honoured mode of self-authorization. *Mandeville's Travels* consists almost entirely of plagiarized passages from other travel accounts, passages cleverly stitched together and rhetorically heightened by claims of personal experience (occasionally leavened by sententious criticisms of the supposed exaggerations and plagiarisms of other, less scrupulous voyagers). Thevet is a master of this genre. To be sure, in his early years he did actually travel and observe and record, but his cosmographical volumes insist on the primacy of experience over authority and claim direct observation even when he is shamelessly engaged in absorbing and reproducing a host of authorities. As we have seen, Thevet's contemporaries were not blind to the game, and the rules were beginning to change. In his great collection of travel documents, *The Principal Navigations, Voyages, Traffiques & Discoveries of the English Nation*, Richard Hakluyt, who had direct dealings with Thevet for several years, prints Thomas Nichols's account of the Canary Islands, occasioned, its author writes, by the 'great untruths, in a booke called The New found world Antarctike set out by a French man called Andrew Thevet'. 'It appeareth by the sayd booke', Nichols remarks sourly, that Thevet 'had read the works of sundry Phylosophers, Astronomers, and Costmographers, whose opinions he gathered together' (VI, 125). Nichols does not intend this characterization as a compliment, but it does fairly describe one major aspect of what Lestringant calls the cosmographer's *atelier*. The competitive world of Renaissance book publishing intensified the claims of individual authorship, but in fact cosmography was necessarily a collaborative performance, the product of a workshop.

Mandeville's fourteenth-century assemblage of texts led in the direction of an unusual tolerance: he praises the Muslim sultan and his well-ordered state, admires the intense piety of Indian idol-worshippers and is led by his encounter with the virtuous Brahmins to declare that 'men should despise no men for the difference of their laws.' For, Mandeville adds, 'we know not whom God loves nor whom He hates; and therefore when I pray for the dead and say my *De Profundis*, I say it for all Christian souls and also for all the souls who need praying for'.[7] By the mid-sixteenth century, knowledge of the world had expanded, along with the power and ambition of European states, but the spirit of tolerance seems to have contracted. 'These wilde men of America', Thevet writes of the Tupinamba among whom he spent ten weeks, 'haue no more ciuilitie in their eating, than in other things.' 'Walking in darknesse, and ignorant of the truthe', they are not 'reasonable' creatures and 'are subiect to many fantastical illusions & persecutions of wicked spirites'; indeed they worship the devil. They war among themselves 'euen like brute beasts' and eat their enemies: 'We finde not in no Historie of any nation, be it neuer so straunge and barbarous, that hathe vsed the like crueltie as these haue done.'[8] The 'poor folk' of Thevet's Antarctica, as Lestringant writes, possessed neither the natural virtue attributed to them by Montaigne nor the simple happiness projected on them by European pastoral or Ovidian poetry; rather they 'defined the contradictory paradigm of a non-Christian humanity'.

But in one of the most interesting surprises in a study filled with unexpected turns, *Mapping the Renaissance World* argues that Thevet's portrait of the Brazilians is in fact considerably less intolerant and reductive than the 'noble savage' of the *philosophes*, 'a pale abstraction fleshed out by no concrete ethnographic content'. The difference comes about not so much because Thevet finds particular features to admire in the very peoples whose barbarousness, ignorance and cruelty he has roundly condemned – though he repeatedly does so – as because all recorded features, whether virtuous or vicious, are for Thevet 'singularities', traits that are irreducible and often contradictory. That is, the very incoherence of his cosmography allows a jumbled confusion of remarkable observations, much like the unsystematized, wildly various contents of the 'wonder cabinets' beloved of Renaissance collectors. Thevet's Tupinamba are like living wonder cabinets: 'Cruel and debauched or virtuous and hospitable, a man of honour or a "great thief", the labels stuck on him in turn or simultaneously seem regulated', Lestringant observes, 'by a constantly mobile code modelled, from detail to detail, on the particularity being thrown into relief on each occasion.' And Thevet himself seems the embodiment of what Claude Lévi-Strauss characterizes as 'la pensée sauvage', the mind of the 'bricoleur' who makes do with whatever tools and materials are at hand, fashioning out of a finite series of elements a whole mythopoetic world.

Hence the big summary judgements, generally coarsely intolerant, at which we have already glanced, function paradoxically not as ethnographic straitjackets but as large, featureless rooms into which Thevet can throw whatever he has picked up from his own observations and from those of his informants (such as mariners, soldiers or the notorious 'truchements Normands', Norman translators who lived with the natives and were said to have stopped short at nothing, including cannibalism), as well as from his reading and imagination. Take, for example, Thevet's flat declaration that the Brazilian savages have no more civility in their eating than in anything else they do. 'As they haue no lawes to take the good, & to eschue the euil', he sententiously declares, 'euen so they eat of al kinds of meats at al times and houres, without any other discretion.' No sooner has Thevet clear-cut the forest in this brutal way than he begins to describe in remarkable detail the actual trees:

> In deed they are of themselues superstitious, they will eat no beast nor fish, that is heauy or slow in going, but of all other light meats in running & flying, as Venison and such like, for because that they haue this opinion, that heauie meates wil hurte and anoy them when they should be assailed of their enimies. Also they wil eate no salte meates, nor yet permit their

children to eate any . . . In their repast they kepe a maruellous silence, the
which is more to be commended, than amongst vs that bable and talke at
our tables, they doe seethe and roast very well their meate, and eat it
measurably and not rashly, mocking vs that deuoure in steade of eating:
they will not drinke when they eate, nor eate when they drink.⁹

On and on the ethnographic particulars pour out, utterly confounding the initial statement – a statement not, however, retracted – that
the savages have no 'discretion' in their eating. Not only does Thevet
hold up certain of these particulars as implicit or explicit reproaches
to European eating habits, but he also offers (apparently at second
hand) one of the earliest instances of the international standard of
taste: the Tupinamba, he reports, eat giant lizards whose meat 'as they
say that haue eate thereof' tastes like chicken.

Where are we at the end of Thevet's account of Tupinamba eating
– or religion or warfare or marriage or any of the other practices that
inspire his blend of bullying, borrowing, inventing, observing,
singularizing? It is difficult to say. Lestringant cannily invokes the
imaginary encyclopaedia by Borges that fascinated Foucault, with its
system of correspondences that allows the most incongruous things
to be conjoined and thereby ruins all principles of systematic classification. Elsewhere, to characterize the twisted, hidden links that the
cosmographer forges between self and Other, savage and civilized,
ancient and modern, Lestringant invokes Deleuze and Guattari's notion
of the rhizome, with its strange, subterranean propagation. But it is
Lestringant's concept of the geographical imagination, a concept in-
fluenced by the ground-breaking theoretical work of Michel de
Certeau, that is at the centre of *Mapping the Renaissance World*. The
book enables us to understand this geographical imagination in its
cultural and historical specificity. And it leads us to understand Thevet
as a strange, compelling instance of the Renaissance fascination with
the invention – at once the finding and the fabricating – of reality.
Spenser responded to reports of 'fruitfullest Virginia' by insisting that
his own fairyland could therefore claim to be real. Thevet actually
glimpsed the New World and returned to tell his royal master and his
countrymen what he had witnessed. Among its singular marvels was
an island 'set as eight degrees ten minutes on the other side of the
Equator to the south-southeast'. On this island, which had never
before been discovered, Thevet landed to find fresh water:

The inhabitants and governors of it were only birds of divers plumages and
sizes in great numbers, but also beautiful fruit trees of several kinds and
colors. And when we thought to go into the isle, because of its thick
woods, I perceived some hills and on these I discovered that there were

some grape leaves. As for the trees I never saw any like them, and you would have judged this place to be a second paradise.[10]

The island, utterly spurious yet carefully drawn in a map as if the cosmographer himself hoped to return there some day, is labelled 'l'Isle de Thevet'.

Stephen Greenblatt
University of California

Notes

1 Claude Lévi-Strauss, *Tristes tropiques*, trans. John Russell (New York: Atheneum, 1970), p. 85.
2 Italo Calvino, *Invisible Cities*, trans. William Weaver (New York: Harcourt Brace, 1974), pp. 106–7.
3 Sir Walter Ralegh, 'The Discoverie of the Large, Rich, and Beautifull Empire of Guiana', in *The Principal Navigations, Voyages, Traffiques & Discoveries of the English Nation*, ed. Richard Hakluyt, 12 vols (Glasgow: James MacLehose & Sons, 1903) X, 349–50.
4 *The Travels of Sir John Mandeville*, trans. C. W. R. D. Moseley (Harmondsworth: Penguin, 1983), p. 137.
5 Ralegh, 'The Discoverie', p. 406.
6 Ibid.
7 *The Travels of Sir John Mandeville*, p. 280.
8 *Les Singularitéz de la France antarctique*, trans. Thomas Hacket (London, 1568), pp. 46, 52–3, 59, 62–3. The title of Thevet's work in Hacket's translation reveals the range of interests to which it appealed: *The Newfound worlde, or Antarctike, wherein is contained wonderful and strange things, as well of humaine creatures, as Beastes, Fishes, Foules, and Serpents, Trees, Plants, Mines of Golde and Siluer; garnished with many learned authorities ... wherein is reformed the errours of the auncient Cosmographers.*
9 Ibid., pp. 46–7.
10 *André Thevet's North America: A Sixteenth-Century View*, ed. and trans. Roger Schlesinger and Arthur P. Stabler (Kingston and Montreal: McGill–Queen's University Press, 1986), p. 50.

Preface to the English language edition

It might seem that, in its translation from French into English, this book has changed its scope, not merely in addressing a wider audience, but also by adopting a more wide-ranging title. *L'Atelier du cosmographe* implied the image of a geographer confined in his workshop; *Mapping the Renaissance World* suggests an opening-up to the distant unknowns of the great voyages. The object obviously remains the same, but the two titles shed a complementary light on it.

The notion of the 'workshop' put the emphasis on the affinity of Renaissance geography with the *cabinets de curiosité*. This aspect is prominent in the work of André Thevet (1516–92), the last Valois kings' cosmographer, who is at the centre of this book. Comparable to an artist's workshop, Thevet's cosmography resembles a jumbled series of unfinished investigations, made up of more or less completed projects and sketches, with a heap of heterogeneous objects, which at first sight seem pointless. It is reminiscent of the different workshops which Picasso used during his long career, in Paris, Boisgeloup, Vallauris and Vauvenargues, and which, after his departure, he always left uncleared, full of an incredible amount of bric-à-brac. Luckily enough, Thevet's cosmographical work in progress has reached us nearly intact, with the creative untidiness of its tools and its materials, both unusual and ridiculous: maps of islands by the hundred, draft copies of his last unfinished books, representing up to four distinct stages of his work, and meticulously annotated mariners' charts. This *Wunderkammer* lacks only the monsters and prodigies that Thevet collected in what he called his 'most precious cabinet'. All this provides an ideal base for an 'archaeological' exploration of Renaissance geography.

The second meaning of 'workshop' is equally relevant. Renaissance cosmography is a collective undertaking, requiring a master with craftsmen under him, in this case disinterested informers, ghost-writers and the Flemish engravers he brought to Paris. Such cosmography is built up on a do-it-yourself basis which exploits many sorts of materials: travel narratives and navigation reports, along with indigenous artefacts and myths. It relies on popular legends as well as scholarly tradition, the latter, paradoxically, being refuted with all the more energy since it provides one of the main threads of the cosmographic whole. After visiting the geographer's working place, this book then turns to the construction of the world.

The English title, *Mapping the Renaissance World*, which was suggested by the publisher and of which I approved immediately, may seem too optimistic, but it provides an accurate expression of the geographical ambition of the Renaissance: the reduction of the world to a wooden sphere, the summation of its burgeoning diversity on vellum paper or on an atlas sheet. More subtly, perhaps, it suggests that cosmography – the science of the world – was not a hierarchized and ordered construction, in spite of its effort to grasp the principle and the totality of the world, but that it is conceived, rather, as a landscape made up of drawing and writing, with its upward and downward strokes, and its reliefs and empty spaces. The reference to the cartographer's activity also suggests the role of aesthetics and play in the inventory of the cosmos, which at the same time is an exploration of the endless archipelago of knowledge. From this derives the double character of the process: looking both backwards to the past and ahead to the future. Through his interest in the 'primitive thought' of the inhabitants of Brazil and his intuition of the incompleteness of the world as it was then known, the 'naïve' Thevet might well be one of the first postmodern writers.

At this point, I would like to thank Michel Jeanneret, one of the very first readers of this research, whose friendly help has opened the way towards this English edition. The translation is due to the competence of David Fausett, who has enriched the notes with explanations that were necessary for an English-speaking public. He is also the origin of some of the remarks on the fine (but unfaithful) translation by Thomas Hacket, who was more generous than subsequent critics in considering 'Master Andrewe Thevet' as an 'excellent learned man'.

Frank Lestringant, 1993

Overture: Renaissance and Cosmography

In approaching the geographical literature of the Renaissance, several critical models come to hand. One might, like Gilbert Chinard early this century,[1] have recourse to the notion of 'exoticism' in order to gauge the evolution, at the progressively enlarged fringes of the *oikoumene*, of marvellous realities bequeathed by earlier ages and gradually idealized or allegorized into new myths. Thus, an enrichment of the stock of prodigies handed down by Pliny and his followers, such as Solinus and Pomponius Mela, had the effect of slowly eroding traditional taxonomic frameworks. From the resulting encyclopaedic chaos (which is well reflected in the work of André Thevet) there belatedly arose that figure of the Other, the Noble Savage, whose euphoric portrait would take two centuries to crystallize. From Columbus to Chateaubriand, by way of Montaigne and Rousseau, one witnesses the painful and constantly delayed birth of that man of nature, each time more youthful and free of servitude.

This positivist vision of history tends to sin by falling into teleological illusion, and can be corrected by means of a second model: that of the 'new horizons' put forward earlier by Geoffroy Atkinson.[2] His paradigm has the merit of privileging – at least in theory – geographical space over chronology, the surface of expansion, by contrast with a linear historical development. This enables one to reach, in principle, the nub of the problem. Thus, Atkinson shows the relatively minor importance of the reception of the discovery of America in the Renaissance, compared to an eastern horizon of expectations whose age-old prestige was actually enhanced at the time by the peaking ascendancy of Ottoman power. The foreign reality that literally 'obsessed' Europe in the time of Suleiman was not that of naked Indian cannibals springing from the depths of the Brazilian jungle; it was that, close to home and yet at the same time more distant, of the Muslim Turk pitching camp and raising his crescent flag on the very doorstep of Christianity.[3]

Such an analysis also falls into error, however, by ignoring the question of scales. The Mediterranean space in which that confrontation between Christianity and Islam was played out – between a Europe torn by religious schism and national rivalries, and the seemingly monolithic empire of the Great Turk – had by this time ceased to be confused with global space. It was thenceforward an arbitrary distortion to place on the same level phenomena belonging to maps – world maps or *mappae mundi*, and chorographies of the Near East and Balkans – whose scales did not coincide. Thus, the author of the 'New Horizons' reproduces an illusion which Renaissance men themselves only reluctantly gave up: that of privileging the Mediterranean centre over an unknown or poorly known periphery, and of forcing fragments of the world that were disproportionate to each other to coexist within the same frame of representation.

Now savants in the sixteenth century – most notably, the historians of Venice, the city located at the very pivot between the two antagonistic cultures – began to realize this disparity of spaces (a disparity that had less to do with their 'quality' than with their 'quantity'), and to take up for the modern world certain categories dear to the cosmographical science that was undergoing vigorous revival at the time. Thus two great collections of historical and juridical documents appeared within a few years of each other in the Most Serene Republic, but subdivided the world differently: not in terms of the fundamental directions of space (east and west, north and south), but according to the distances and orders of magnitude they envisaged. Between Giovanni Battista Ramusio and Francesco Sansovino, the editors respectively of *Navigationi e viaggi* (1550–9) and the collection *Dell'historia universale de' Turchi* (1560), the boundary is to be traced not in terms of meridians, nor of any geographical lines whatsoever; since the 'Orient' is the object common to both enterprises. It is *scale* that forms the division between these two complementary collections.

Ramusio is interested in the Far East, and makes his domain the distant outside world – that of 'navigations and voyages', as his title explicitly indicates; whereas Sansovino devotes his documentary collection to an intermediary region, Turkey and Persia, which is posited as the precise negation of the Christian West. The distinction between the two might, therefore, seem to be one of genre, as Stéphane Yérasimos has suggested[4] – distant peregrinations being opposed to a history primarily concerned with the immediate neighbourhood, to which peoples more recently discovered had no right of access. One might, too, provisionally agree with Yérasimos, that in the outline of the world offered by European humanism in the second half of the century, 'the more geography there was, the less history.' But this apparent distinction obscures another that, in my view, is more essential: the small scale of global representation was radically distinct from the medium or large scale appropriate to a region, be it more or less extended, of the earth. The former grasped the *quantity* of the world, whereas the latter plumbed its *quality*. A planisphere that reduced the terraqueous globe to its broad outlines did not admit of the same objects as a partial (chorographic or topographic) map swarming with a profusion of different places. A history of events, right down to the cycle of seasons, could easily enter into the latter type of map by way of a large qualitative scale that allowed one to fix accidental details, to inscribe locally the passage of the present. Thus, the gold of harvests or enamelled prairies of flowers formed part of the programme that Girolamo Cardano prescribed for the perfect chorography.[5]

On the other hand, the small scale of the *mappa mundi* lent itself ideally, in a future-oriented vein, to audacious strategic anticipations. The reduced scale of cosmography, or universal geography, seemed ideally suited both to the dreams of the navigator and to the speculations of princes or diplomats. To them it was given to 'sculpture the azure ocean'[6] – to carve in it, compass and dividers in hand, the boundaries of purely theoretical spheres of influence. In this sense, the Treaty of Tordesillas might be considered the first cosmographical act of the Renaissance. Concluded on 7 July 1494 between Portugal and Spain, and ratified by Isabel of Castille on 2 August and by John II of Portugal on 5 September, it rigidly divided the two empires by means of a meridian or 'direct line traced from pole to pole', set 370 leagues west of the Azores.[7]

For cosmography did not allow itself to be encumbered by obstacles. From the lofty position it took up it effaced all relief, and abolished every feature of the land. Indeed, its privileged field of action was doubtless that constituted by the vague and unified expanses of the oceans. But to the real configuration of the globe it was, one might say, indifferent. Given that 'it divided the world according to the circles of the heavens',[8] and that its lines of force resulted from a projection on to the sphere of the circular movement of the stars (within, of course, the geocentric system of Ptolemy), cosmography could reign as an absolute sovereign over the terraqueous globe. It manipulated at will the natural frontiers of rivers and mountains; determined the futures of peoples by fixing their migrations and boundaries; remodelled, if necessary, the structure of continents; and controlled the calculated drift of archipelagos.[9]

By virtue of this future-oriented dynamism ruling over an unfinished present, cosmography was diametrically opposed to the regional detail of chorography. The latter recorded from place to place the events of the past, and made the regional map into a genuine 'art of memory' in the sense that classical antiquity attached to the term.[10]

The topographer's landscape-map was a profuse and indefinitely fragmented receptacle of local legends and traditions that were rooted in vagaries of relief, hidden in folds of terrain, and readable in toponymy and folklore; whereas the reticular and geometrical map of the cosmographer anticipated the conquests and 'discoveries' of the modern age. No doubt the marvellous was not absent from it; but it subsisted there only by special dispensation.

If, for example, the Le Havre pilot Guillaume Le Testu placed at the margins of the known world, in his *Cosmographie universelle* of 1556, monstrous populations inherited from Pliny, St Augustine and Isidore of Seville, it was only in order to establish provisional boundaries for a knowledge in a perpetual state of progress.[11] 'Progress' means here that enlargement of a space that was pushing out on all sides and stitching together, as voyages allowed, the remaining gaps in it; rather than the linear and continuous development of a rectilinear history of knowledge.[12] There was not such a great difference between the *Monoculi* (one-eyed men), Sciopods, Dog-heads or other *Blemmyae* that haunted the depths of Asia (but also reared their heads elsewhere: in the most impenetrable regions of the New World, or the fabled Southern Land), and the *padrões* or milestones erected by Portuguese navigators here and there on the shore to mark their progress along the African coast; for on such milestones, painted in the colours of fables, also rested the future advance of a great quest.

Having failed to take account of the differences of scale between traditional chorography and the Ptolemaic cosmography renewed by Münster and his followers, we have tended to assume an imbalance of Renaissance 'geographical literature' in the Orient's favour. The result has been to confuse different orders of magnitude and, with them, spaces and objects calling for different methods of analysis.

A second error has flowed from this. Having failed to recognize the 'cosmographic revolution' taking place at the turn of the sixteenth century – the sudden rupture of scales that changed people's way of viewing the world, and consequently the world itself – criticism has often remained trapped in a narrowly historicist vision of travel literature. This has resulted in an undue eagerness to sort out 'good' from 'bad' geographers, to distinguish the 'forward-thinking' from the 'backward'.[13]

Now the chorus of insults that greeted a cosmographer like Thevet during his lifetime, as the result of a cleverly orchestrated polemic,[14] mostly did not emanate from the more 'progressive' Renaissance writers. We find among the enemies of cosmography, alongside undoubted representatives of the 'new historicism' (sixteenth-century style) such as Urbain Chauveton or Lancelot Voisin de la Popelinière,[15] the most rigorous exponents of the theological tradition it threatened. The Catholic 'zealots' Gilbert Génébrard and François de Belleforest, but also the Lutheran Ludwig Camerarius, were scandalized less by the medieval candours of Henry III's cosmographer than by the blasphemous audacity of his undertaking.

Beyond the casual alliance his adversaries formed against him, therefore, one can ask if Thevet, far from having sinned against the truth and against his century,[16] was not on the contrary being taxed with its most intrepid innovations. A case in point is the Brazilian 'Henryville' that Huguenots and Leaguers hotly disputed, but which was in the end nothing but a strategic fiction, a poorly understood colonial anticipation.[17] A similar misunderstanding arises in connection with Thevet's curiosity about New World cultures, his drawn-out transcription of Tupinamba cosmographical myths or his use of an Aztec codex from the beginning of the conquest period for the Amerindian iconography of his *Vrais Pourtraits et Vies des hommes illustres*.[18]

The pride and excess that his contemporaries ceaselessly stigmatized in Thevet appear, in fact, to be part and parcel of the cosmographical project. From the part to the whole, and from the eyes and ears of the world to its face (to use an ancient 'similitude' illustrated by Peter Apian: see plate 2), the leap from partial chorography to global cosmography involved an upward displacement of one's point of view.[19] It is incorrect to say that the observer's view expanded; and in this sense, too, the concept of 'new horizons' seems somewhat imprecise. The point of view was elevated, to the point of grasping in a single instant the convexity of the terraqueous globe. At that imaginary point, the eye of the cosmographer ideally coincided with that of the Creator. Spatial hyperbole allowed this passage from the closeted world of chorography to the plenitude of a universe revealed at last in its totality.

Such a leap from the qualitative to the quantitative, and from the earth to the heavens, was not accomplished without some difficulties. A chasm lay in the path of the cosmographer who would transform himself into Lucian's Icaromenippus. How might he embrace this totality, at the same time overflowing and lacunary, of the cosmos? How could he bridge the gap between a theoretical global vision and the millimetric apprehension of singularities that 'natural philosophers', as faithful inheritors of the medieval *Imagines mundi*, obstinately continued to gather? Furthermore (and this problem became acute during the Counter-Reformation) it was difficult for a cosmographer to escape the accusation of pride in so far as he pretended, as his profession required, to embrace with his vision and, as it were, to grasp in his hand the two extremities of the theatre of Nature: the local scale of individual experience, and the universal scale of the divine plan.

A formal submission to theology, reiterated at the beginning of each of his books, allowed Thevet to escape the suspicion of heresy which fell, by contrast, on Guillaume Postel, his friend and companion in the Levant. This caution, coupled no doubt with a basic inability to raise himself to the heights of abstraction or to penetrate the arcana of nature, caused him to despise the speculative sciences. As a pupil of the geographer Oronce Finé (Orontius Finaeus) and a friend of Antoine Mizauld and of Postel, he ignored the astrological profession, and had nothing but contempt for adepts of the Cabbala, even the Christian one.[20] Similarly, two chapters of his *Cosmographie universelle* of 1575 sufficed to dispatch the mathematical aspect of the programme he was committed to, and he would later return to it only surreptitiously, in pages whose incoherence has been emphasized.

Being tainted with hubris, the cosmographical enterprise did not serve the aims of natural theology at all well. It would take all the efforts of Sebastian Münster and Richard Hakluyt, within the perspectives of the German and Anglican Reformations; and, later, those of Mercator and Hondius in the Low Countries of the Catholic Reconquista, to christianize a discipline that sinned by its excess of self-confidence. Only then would be affirmed the fecund tradition of 'cosmographical meditations' that, from Vadianus to Mercator, made the contemplation of the atlas one of the privileged means of access to an understanding of the Scriptures.

'From the Mappe he brought me to the Bible.'[21] With this startling recollection, Hakluyt describes his conversion to geography under the aegis of his cousin Richard Hakluyt the Elder, who, receiving him one day in his study, guided his reading from the planisphere to the Psalms. The commentary on a map was a spiritual exercise like any other, and it offered as well the advantage of not separating the believer's interior reflections from his or her practical activity in the world. The admiration aroused by the spectacle of the Creation miniaturized into a map went hand in hand with an examination of the 'commodities' the geographical apprentice could find there. The beauty of the cosmos thus resided in its use value, and in the profit that a Christian could draw from it. Spontaneously the young Hakluyt discovered that the service of that generous God whose humble chaplain he was to be, and the design of a greater England, were the two faces of a single duty.

But there was no such liaison for Thevet. The dreams of colonial implantation that he nurtured for the France of the last Valois kings, and that he situated by turns in the Bay of Rio de Janeiro or in the St Lawrence estuary, did not at all correspond to a belief in the providential destiny of a nation. More simply, in creating forts and crenellated citadels at random on formerly savage shores he was fulfilling the offices of a dutiful courtier. His exercise of a fantastic geography pursued no other end than the Prince's contentment, and secondarily also his own megalomaniac tastes as the 'Cosmographer to four kings'. For want of any higher ambition or of the transcendental necessity that pervades the writings of Hakluyt, an Anglican chaplain who became a propagandist for the colonialist and Puritan lobbies in Paris and London,[22] Thevet's cosmographical fictions would remain of no political consequence. An age dogged by misfortunes and persistent civil war encountered there a dream reduced in advance to nothing by its sheer gratuitousness.

Furthermore, unlike Münster or even Mercator, who opened their *Cosmography* and atlas respectively with a Creation-story inspired by the biblical Genesis, Thevet broke with this tradition of marrying a profane geography to sacred history, and chose to begin *ex abrupto* by recalling the definitions of Ptolemy.[23] Far from celebrating the wedding of Moses and Cosmography (like Martianus Capella celebrating that of Mercury and Philology), his work endlessly announces a divorce between the excesses inherent in the programme of Ptolemy and the Alexandrian school and the humility required of a Christian philosopher. Whence Thevet's 'blasphemies' in relation to the Old Testament, of which he imprudently rejects fables such as Jonah's whale, Samson's lion or Ezekiel's Pygmies.

The problem posed by cosmographical hyperbole was not only one of theological doctrine; it bore on method itself. Cosmography, whose rehabilitation coincided with the great discoveries, paradoxically developed at a time when the new state of the world should have shunted it into obsolescence. The earth had grown in spite of it, and the old *oikoumene* that the ancients had limited to a longitudinal portion of the northern hemisphere was multiplied by four.

Yet this model, apparently so inadequate, proved to be fecund by virtue of its very anachronism. It offered to modern geographers a three-quarters empty canvas, leaving them free to inscribe on it the delineation of newly 'invented' or discovered lands; a form, at once closed and open, full and lacunary, that represented the ideal construction in which to house, with their approximative and disparate localizations, the 'bits' of space that navigators brought back from their distant voyages, having summarily entered them in their rutters

or portolan charts.[24] As Jean Lafond notes, 'a good model was pro-
ductive in so far as it could be applied to another domain than that
with which it was hitherto associated.'[25]

In like manner, one could argue that the continuing productivity of
Ptolemaic cosmography in the Renaissance arose from the fact that
the world-object was no longer what it had been in late antiquity. In
relation to the reality gradually emerging from navigation on the high
seas, the cosmographical model seems at once anticipatory and backward. It was backward on account of the limits it fixed for human
curiosity and action, which one by one would have to be overthrown.
Experience would reveal that the uninhabitability of the glacial and
torrid zones was a doctrine without foundation; there was no exact
symmetry between the northern and southern hemispheres, and
America, drawn out in latitude, contradicted the general organization
of the Old World, its division by the Mediterranean into the three
riverine continents of Europe, Asia and Africa. As for the Aristo-
telian doctrine of progressively encased spheres of earth and water, it
posed insoluble problems for the cartographer and would soon have
to give way to the modern conception of a terraqueous globe.[26]

But it was a model that could be corrected: it was perfectible, to the
extent that it largely anticipated the new state of practical knowledge.
Virgin spaces subsisted on the sphere; notably around the poles, where
they aroused endless speculation about the existence of a north-west
passage or a southern continent. The depths of the North American
or African interiors also long escaped the general enlargement of
perspective and the growing extension of global economic fluxes. Thus
the universe, while theoretically full, remained in practice malleable.

The inadaptation of the cosmographical model was fecund in a
double and contradictory way: being at the same time experimentally
backward and anticipatory in mathematical terms, it opened to Renaissance science a space in which could be played out the various
national and personal projects or 'cosmographical fictions' referred to
above. Thus the England of Ralegh and Hakluyt invented for itself a
northern empire that would take half a century to realize. As for the
French monarchy, it found itself endowed with an erratic New France
whose basis moved, in the course of decades, from Brazil to Florida
and Canada; it even came to rest for a while in the hypothetical
southern land, which La Popelinière claimed, at the time of the Portuguese war of succession, for the declining Valois dynasty.

Thevet, for his part, would exploit that model; its ambitious framework, openness in principle, and internal play would leave him free
to dispose of the profuse diversity of the world as he pleased. The
share of initiative that devolved on to the cosmographer in the exercise

of his profession was immense, though not unlimited. In effect, the prestige that it accorded to extension freed him from the old servitude to chronology. Thevet played on it (unduly, no doubt) in order to sweep away the scientific tradition whose inheritor he claimed – contradictorily – to be.

Cosmography, it will be clear, was also the project of a lifetime commitment. Before setting about an examination of Thevet's work, it is worth sketching the broad outlines of a career that was in many ways exemplary.

He was of modest origins, a younger son of a family of surgeon-barbers in Angoulême. His parents placed him in the Franciscan convent in that town at the age of ten; and it was his religious order that would allow him to travel and accede *de facto* to the discipline of geography. A first 'periplus' took him to the Levant from 1549 to 1552. Made a Knight of the Holy Sepulchre in Jerusalem, he had to fulfil, along with his itinerary of pilgrimage, some mission of a diplomatic nature that retained him for two years in Constantinople. On returning to his convent he supervised the writing of a *Cosmographie de Levant*, which however owed more to the work of compilation by humanist authors than to his own memories. This work, abundantly illustrated, was published in Lyon in 1554.

The real springboard for his career as cosmographer to the kings of France was offered by a second voyage, this time to the New World, in the company of Nicolas Durand de Villegagnon, Knight of Malta.[27] This adventure in 'Antarctic France' – that exiguous austral France founded on an islet in the bay of Rio de Janeiro – was for Thevet limited to a brief winter spent 'among the most savage men of the universe', from 15 November 1555 to 31 January 1556. Falling ill shortly after disembarking, he was sent home on the ship that had brought him. From this short season in the tropics, however, he would draw the material for a second book, much more topical than the previous one and destined finally to establish his fame. The *Singularitez de la France Antarctique*, published at the end of 1557, was translated into Italian and English, and for a long time the work was the subject of borrowings, imitations and polemical debate. The quality of its documentation of the flora, fauna and manners of the Indians of southern Brazil, and its rich illustrations, in which fantastic scenes decked out with native artefacts sat beside botanical plates devoted to the manioc, banana or pineapple plants – all assured the work a wide circulation at court and among poets and seekers of curiosities. Its depiction of an Edenic Brazil stimulated the literary rivalry of Jodelle and Jean Dorat or of Du Bellay and Ronsard, who tried to outdo each other in composing original variations on the conquests of Jason and the theme of a lost golden age and the naked liberty of our earliest forefathers.

It was following this publication that Thevet was able to realize the great ambition of his life. Released from his vows at his own request in 1559, he soon became chamberlain and then cosmographer to the king. That office seems not to have existed in France before him; perhaps he himself created it, following Spanish and Portuguese models. Its functions are vague, and its level of remuneration uncertain: Thevet would never be one of the experts in nautical science that the sovereigns of the Iberian peninsula so prized, nor a holder of state secrets such as would have made him equivalent to a minister or privy counsellor to the prince. There was in France, throughout this whole period, no coherent maritime policy; initiatives in this area came and went without any order, emanating by turns from the various hostile factions that fought for the king's ear – sometimes from Coligny, sometimes from the Guises. Thevet's role therefore remained, it would seem, of secondary importance. Hence the discredit that befell him even during his own lifetime, and the almost complete success of the 'cabal of the learned' that was formed, from the 1570s, against this autodidact who was so insolent as to pretend to a monopoly of geographical knowledge in the French kingdom.

From his association with the powerful, and notably with the Florentine entourage of Catherine de Médicis, he would nevertheless gain secret information that had some strategic value: such as that issuing from the *Histoire notable de la Floride* (1566) by René de Laudonnière, or from the memoirs of Roberval concerning a colonial establishment in Canada.[28] But from these documents he drew nothing more than the material for impenetrable chapters that form the substance of his last works, left in manuscript: the *Histoire de deux voyages aux Indes australes et occidentales*, and especially the *Grand Insulaire et Pilotage*, an atlas with commentaries containing more than 300 charts of islands and islets all around the world.[29]

The fact remains that, from this time on, Thevet professed ambitions on a universal scale. It was then, once established in Paris around 1560, that he began the most controversial phase of his career and his work. His image had hitherto been that of a long-distance traveller; it was as such that his contemporaries hailed him, knew him, and sang his praises in dithyrambic odes. But to the qualities of endurance, courage and curiosity that such a role required he now claimed to join those of unlimited knowledge and perspicacity. In this new guise he courted the favours of the great, played off Catholic and Protestant noblemen, and directed from a distance an ever more vast enterprise of compilation. His contribution to the latter consisted mainly of pouring in documents that were inaccessible to anyone but himself, and of which he was often the only person to perceive the interest. Thus his collection of Americana, in which appear, beside the *Codex Mendoza*, fragments on the religion of the Tupinamba and the Aztecs, nourished the fourth volume of his *Cosmographie universelle* in 1575, and a decade later colonized the eighth part of his *Vrais Pourtraits et Vies des hommes illustres*. The latter opened up to heroes of the New World the gallery of personalities bequeathed by Plutarch, which was thereby enriched by the great men of his own age.

At the time when the principle of collections of voyages and related documents triumphed, cosmography became more and more obsolescent. But it remains the only model that allows us to link together the two divergent periods of Thevet's life, and the two radically distinct strata of his work. Through it, we cease to see the tearaway experiences of his youth and his sedentary old age as contradicting each other. His accounts of his voyages to the Levant or among the man-eaters of Brazil structure the immense compilation that progressively sedimented around those two original tropisms. In order to effect a fusion between the observed and the borrowed, cosmographical fiction as practised by Thevet promoted certain dominant ideas or, if one prefers, obsessional themes: the primacy of experience over authority; the sovereignty of a universalist view enveloping in an instant the terraqueous globe; and a preference among sources for the technical and 'popular' writings of pilots and mariners.

Such a project, no doubt, defined no method worthy of the name. But it bore witness, before the durable compromise that the classical age would impose between tradition and novelty, to the extent of the crisis of cosmography; one that for its part it tried to remedy using the means that came to hand, and not without a certain brutality.

The enterprise ended in an impasse; Thevet would have no immediate posterity. For two centuries he would function as a whipping boy for savants infatuated with clarity and order, and moreover indifferent to the radical foreignness of ethnological singularities brought back from distant horizons. Yet in the extreme attention it paid to *realia*, in the secret affinity for the 'savage mind' of the Indians that it outlines with rare precision, Thevet's work (itself arising in a time of crisis) announced many of the preoccupations that are ours today. He was one of the first to treat with suspicion, with a gaze obstinately applied to the diversity of things, the human universalism that would have the force of law throughout the classical age.

1

The Cosmographical Model

I would ask of those who held such an opinion: even if they had studied books of cosmography and navigation for fifty years, and had maps of all regions and dials, compasses and other astronomical instruments, would they yet undertake to steer a ship to any land? Would they not, like any trained and experienced man, be wary of placing themselves in such danger, however much theory they had learnt?

Bernard Palissy, *Discours admirables de la nature des eaux et fonteines*

... in these matters, the most savant see less clearly than do the sailors and others who have long since travelled to those lands; for in all things experience is our mistress.

Thevet, *Cosmographie universelle*

An all-out cosmography

In choosing the cosmographical paradigm, as he did from his first published work in 1554, André Thevet – appearances to the contrary – turned his back on the Middle Ages. He was taking up an ancient model that the Renaissance, notably with Sebastian Münster in his role as editor of Ptolemy, had just rehabilitated. The hypotheses of cosmography supposed a full, global world with no other limits than the celestial orb that, projected on to it, formed its poles, regions and zones. This global and geometrical vision of the earth, which Montaigne held in suspicion and which Belleforest freely taxed with blasphemous pride, was derived from a literal reading of Ptolemy, the first chapter of whose *Geography* forms the preamble to Thevet's *Cosmographie universelle*, just as it had earlier served to introduce Sebastian Münster's *Cosmographia*.

Although Thevet felt no particular deference towards the great ancients, of whom he had only a superficial knowledge and whom he began to read only through the compilers of late antiquity, from Pomponius Mela to Solinus, he needed that formal framework in order to institute his own descriptive project and at the same time to give it the authority and breadth it needed.[1] The reference to Ptolemy played, here, the same role as the ubiquitous mention of Pliny in the *Historia general y natural de las Indias* by Gonzalo Fernandez de Oviedo. As is well known, that Spanish chronicler owed to the Latin naturalist, whom he greatly admired, a framework of expression which he rigorously imitated – to the point of making his introduction or 'proem' the first book of his *History*, in the manner of his forebear, the second book forming the true beginning of the narration, with the inaugural voyage of Christopher Columbus to the Indies.[2]

More casual than Oviedo, and less strictly formalist, Thevet had in common with him the fact that they both supplemented the faithful reproduction of an age-old model with its correction in the light of modern experience. With them the practice of 'autopsy' was fixed in the immutable form inherited from antiquity, but from the outset revealed the inadequacy of that ancient framework. The matter was fresh, though cast in an old mould; as Oviedo put it, 'The history I write will, however, be true and abstracted from all fables and false-hoods.'[3] This was because his duplication of Pliny's work in a New World context arose from the author's personal experience. His American collection had been gathered 'through ten thousand labours, necessities and dangers', endured 'for twenty-two years already, and more'.[4]

Before Thevet, then, the conquistador Oviedo had drawn his authority from a long series of physical and moral trials that ensured the veracity of his word. Totally unlike those armchair geographers Aristotle, Ptolemy and Pliny, Oviedo had actually confronted, before writing, the rebellious nature of America; its stigmata remained fixed in his flesh, and were thus reconstituted into a painful and martyriz-ing memory. It was no doubt necessary that such experience of 'the rudeness of the country, its air, the great thickness of its grasses and bushes, the dangers of its rivers, its huge lizards and tigers, the sam-pling of its waters and foods, was at the risk of our lives',[5] in order that merchants and settlers might benefit, 'with cloths laid out and no effort', from the sweat and blood of the first conquerors. But such efforts and labours founded more than towns and colonies that quickly prospered: they grounded the authority of a history written from experience. Not only did the 'eye write', to use a phrase François Hartog proposes to define autopsy,[6] but the voyager's whole body, stigmatized in the course of his long travels, bore a sort of guarantee of the truth of his testimony.

In the same way, Thevet in the course of his work records the numerous qualifying trials in which he constantly seemed to be 'in danger of his person', as he meticulously recalls in the alphabetical tables of his *Cosmographie*.[7] From Jerusalem to Guanabara and from Gaza to Seville, his impenitent curiosity and his duty as a natural philosopher exposed him to a thousand dangers: insults and beatings, blows and wounds, interrogations under torture and imprisonments mark the progress of his painful odyssey. He escapes from Turkish jails only to fall into those of the Spanish Inquisition, and often owes his life to nothing more than his sang-froid, backed up at appropriate moments by divine Providence. His continuous perils, which become somewhat repetitive in this theatre of the four continents, were destined none the less to prove the eminent merits of the author and enable his credit to carry the day over the chilly science of the academic geographers. His prefatory address to Henri III is a precise echo, in this respect, of the preface to the *Historia general y natural* in which Oviedo addressed Charles V. Reminding his patron in turn of the usefulness of geography in affairs of state and enterprises of conquest, Thevet concludes by evoking his services 'in the four parts of the world' and the endless hardships he had endured at sea, exposed as he was 'to an element most inconstant among all others, and at the mercy of winds, storms, tempests, the barbarity and cruelty of foreign peoples, and to an infinity of other perils in which one can hope for death rather than life'.[8] If we are to believe him, then, it was not Thevet's fault that he was not among those empire-builders of whom Oviedo considered himself one of the most authentic and disinterested specimens. Historical circumstances in France lent themselves to little but a semblance of the maritime heroism in which this cosmographer of monarchs was soon to clothe himself – monarchs as transient as they were indifferent to colonial adventure.

The oceanic spaces involved in this martyrdom of the long-distance traveller ensured the link – a problematical one if ever there was – between a personal experience necessarily limited in space and time, and the general consideration of the world that mathematical instruments of cosmography allowed. The sea – that consummately fickle element, distrusted since antiquity – was also, paradoxically, the simultaneously undifferentiated and unified space in which theory could most precisely espouse the practical. On the uninterrupted surface of the oceans, the lines of the heavens could be ideally propagated without encountering any obstacle. The Treaty of Tordesillas in 1494 ratified in its way this privilege devolving on to the ocean, in its simplifying subdivision of areas of influence between Spain and Portugal.

The half-meridian or 'direct line' traced 'from pole to pole' 370 leagues west of the Azores was a line of demarcation that traversed the Atlantic[9] and allowed Portuguese fleets bound for the east coast of Africa and India to follow the circuit dictated by the patterns of trade winds in the southern hemisphere.[10] Driven by necessity and the will of the South Atlantic *volta* (the circuitous route dictated by trade winds), Cabral discovered Brazil in 1500 and made of this line of demarcation a terrestrial frontier, albeit approximate and theoretical;

seizing from the cone of Spanish America a part of its Atlantic façade. But the half-meridian fixed at the time of the Tordesillas conference, which would subsequently have to be completed in order to settle the difficult question of the ownership of the Moluccas, was in the beginning freely traced across the indefinite space of the seas.

It is an abstract place, the ocean, without shape or points of reference. As the pilot Pedro de Medina noted in his *Arte de navegar*, a famous manual used in every port of Europe, it was indeed a great mystery 'that a man with a compass and rhumb lines can encompass and navigate the entire world'.[11] That none the less was the wager offered by nautical cosmography, which would guide the mariner across 'a thing so vague and spacious as the sea, where there is neither path nor trace'. 'And', adds Pedro de Medina, borrowing the authority of the Book of Wisdom, 'to tell the truth, it is a subtle and difficult thing, well considered by Solomon when he said that one of the most difficult things to find is the path of a ship at sea. For it follows no path, and leaves no signs.'[12]

Now it was in this abstract place, the 'proper place of the waters', devoid of relief or definite colour[13] and without any boundaries or routes, that – paradoxically – cosmographical theory and the concrete experience of the navigator coincided. Lashed by spray and salt, thrown about by the swell, a plaything of the unleashed elements, the body of the pilot occupied that position of ideal mastery enjoyed by the cosmographer, between the sea and the heavens. From the castle of his caravel he dominated the horizontal expanse of ocean and contemplated, above him, the passage of the stars that told him his position and route. The enthusiasm or even hubris of the cosmographer also made themselves felt – if we are to believe Thevet – in this geometrical and indeterminate place where the world, reduced to its essential lineaments, appears totally *comprehensible* (in the fullest sense of the word). Such euphoria irradiates, in places, Thevet's *Cosmographie universelle*: in pages where the author pictures himself not as sitting in front of a *mappa mundi*, but as actually inside the latter, at the same time integrated into the oceanic surface that he traverses in his ship and mastering it with his gaze:

> All the things I am narrating and reciting to you are not to be learnt in the schools of Paris, nor in any university of Europe whatsoever; but in the seat of a ship, under the tutelage of the winds; where your pen is the quadrant and compass, and you hold up your astrolabe to the clear light of the sun.[14]

The admirable metaphor drawn from this school of navigation ends up as the living emblem of a sovereign and all-embracing practical knowledge. All mediation is suppressed, beyond indispensable technical tools such as the quadrant, compass or astrolabe; so that the plane surface of the world and human knowledge about the world coincide exactly. The grandiose and proud posture of our travelling cosmographer, disdaining his study-bound colleagues, links up with the nautical apotheosis of Magellan, as represented by Hans Stradan in an engraving published in 1594 in the fourth part of Theodore de Bry's *America*, his collection of 'Great Voyages'. Sitting in his armour on the bridge of his caravel as it passes through the eponymous strait, flanked by an Apollo Citheroedus floating in the air and an arrow-eating Patagonian giant who sits enthroned with his feet muffled up on the adjacent shore, the discoverer holds in his hand the compass he is using to record measurements of angles on the armillary sphere standing in front of him.[15]

Scattered about the bridge and projecting from the sides of the ship are culverins and cannon, indicating that cosmographical calculation – even if it allowed a simultaneous grasp of space – was not, for all that, entirely disinterested. On the contrary, its aim was immediately political. Thevet's enterprise would be, after that of so many others, to transform the intellectual and symbolic possession of the world into a military conquest of it. The fall of the Valois monarchy, the reticence or the whims of successive monarchs, the carelessness of warlords and the weakness of a diversely composed and poorly organized navy would shunt that dream of empire into the realms of fiction. However, our cosmographer did not bear alone the responsibility for the perversion of science into legend through its lack of a field of application or of a means of being acted on.

Returning now to the favourite place of our geographer of the high seas – to that ideal, mobile and radiant point from which he saw both heaven and earth, imagining the one through the other from the unstable bridge of a high-sided ship, a sort of oceanic utopia – we find that Thevet took literally the dream of ubiquity and omnipotence allegorized by Stradan's engraving. A particularly revealing description shows the cosmographer crossing the equator and discovering, within the space of a few hours, the constellations of the two hemispheres. In that magic star-spangled night, in which the heavens were turned around over his head, the voyager could envisage his knowledge expanding progressively to the limits of the cosmos, reducing to vain speculation the science of all his predecessors since the beginning of the world. A Thevetian version of 'Scipio's Dream', one might say; yet one in which the cosmographer's mind is by no means torn from his body or the earth. The ubiquity of his gaze is realized at the human level, and his omniscience is developed in the wake of the caravel that slides along before a trade wind.

What did hunger, thirst and torrid heat matter then, since such delights were reserved only for the erstwhile cosmographer?

> But to have knowledge of rare and excellent things the curious man, as I was, is not bothered by the troubles and irritations that might be visited upon him, because his satisfaction makes him forget the burden of his labours. There it was, that in the space of a few hours I saw the two poles. God had thereby made me happier than Aristotle, Plato, Pliny or others who undertook to speak of the celestial bodies: for what they said about them was only the result of imagination, yet it had been made accessible to me, and opened up to my view. Thus we can now see rising and setting all the stars; not only the two Bears of the North Pole, but as well those that accompany and surround the Austral Pole.[16]

Gilbert Chinard, praising the 'lyricism' of this page (and even more, the suggestive atmosphere of the accompanying engraving, which had already been savoured by Flaubert in his time),[17] reads in it the vibrant memory of tropical nights and their stifling heat.[18] No doubt, considering the greed for picturesque detail that was common among Renaissance geographers and travellers, this evocation is notable for its personal and concrete tonality. Its recourse to the sensual – which is too rare, at least before Léry, not to be remarkable – is none the less in conformity with a definite ideological programme. But the euphoria Thevet felt in crossing the equator was not only of an aesthetic or sensory order. The event demonstrated the double superiority of practice over theory and of the moderns, tracking through a sea at last free and open, over the ancients, who continued to speculate from the confines of their study, and whose gaze rarely saw beyond the regular but limited orderliness of their porticoes. Thevet sees directly with his eyes 'what man believes he sees';[19] what the philosophers of antiquity only imagined by conjecture. His 'happiness' goes well beyond the intoxication of distant navigations, to express the pride of an experimenter who has succeeded in placing the entire world beneath his magnifying glass.

The engraving itself, which shows a ship sailing on a calm night with the wind abaft, gives further proof of this triumphant practice. The pilot, with his cross-staff and magician's wand, standing on the poop of the caravel to measure the altitude of the stars, proves once more the privilege accorded by Thevet to the exercise of what Lucien Febvre calls 'open air geography'.[20] Real space is deployed, without limits, around the observer: a nautical and celestial chart on the scale of one to one, continuously revealed by the ship's regular course to the eyes of a wanderer who is not so much amazed as attentive.

We therefore need to place such a vision – an admirable one, no doubt, but strongly imprinted with the polemical spirit – in the context of a Manichean opposition between practice and theory: those hostile twin personifications who, in Bernard Palissy's somewhat later *Discours de la nature des eaux et fonteines*, endlessly debate the most varied questions of natural philosophy, with the former always emerging the victor.[21] Palissy revealed himself, in this, a follower of French disciples of Paracelsus such as Jacques Gohory or Alexandre de la Tourette; and one might say more generally that he situated himself within an alchemical tradition for which the notion of experiment, with its avatars, 'essay' or 'assay' (see note 24) and 'prove', enjoyed undeniable prestige from the middle of the century.[22] Such experimentalism announced only very distantly the method of Claude Bernard. It was based on fragmented elements of observation gathered by various and isolated practitioners, and this gave it the authority to ruin, by an abusive generalization, the age-old certitudes of the learned. It was, in a sense, a claim for proletarian knowledge; and often emanated from autodidacts and practical workers (thus, for example, Paré was a surgeon, Palissy a potter, Thevet an ill-formed Franciscan). It sought to break down the institutional science of humanists who corresponded in Latin, read Greek and sometimes Hebrew, and proudly ignored those who had not studied as they had.

This parallel with the great navigations was, moreover, also developed by Palissy in order to demonstrate that all the theory in the world was worth nothing without the assiduous exercise of practice. Such comparisons with the undeniable progress observed in the geographical domain were well on the way to becoming a leit-motiv in the empirical literature. Witness, from 1549, the preface Antoine du Moulin composed for his translation of a work on 'Chiromancy and Physiognomy' by Jean d'Indagine.[23] This apology for sciences as inexact as chiromancy, geomancy or physiognomy took the unusual detour of a passage to the New World and the Antipodes. Thus the future triumph of Jean d'Indagine and his French admirers is presaged by the stunning victories of Columbus and Magellan over Ptolemy and Lactantius, those academic speculators who were more at home among climatic zones and angels than on the deck of a ship.

In an intellectual context such as this, it is hardly surprising that personal vision should be the argument of ultimate authority that Thevet opposed to both his predecessors and his contemporary detractors. As far as the cosmographical discipline itself was concerned, one could say, in a pastiche of Montaigne, that this was the 'mystical foundation of his authority';[24] indeed one might add, again with the author of the *Essais*, that it 'has no other'. The primary definition of cosmography, as given in 1564 by the Bolognese Leonardo Fioravanti, in fact makes the whole validity of the discipline depend on experience. This truth seemed so evident to the compiler of the 'Universal

Mirror of the Arts and Sciences' that he did not bother to demonstrate it: 'Cosmography is a science that no man has ever been able to learn or know other than by experience: a fact that is most manifest, and has no need of proof.'[25]

As an inheritor of this self-evidence, Thevet also appropriated its exclusive benefits. From the outset, his personal experience establishes his tyrannical and discretionary power and consigns all previous opinions to oblivion:

> I can indeed say that I have observed certain fixed stars of the Austral land, and that even if I were to listen for ten years to some learned doctor tormenting himself over an astrolabe or a globe, I could have no greater knowledge of the matter... If the Ancients too had seen and recognized them as I have, they would not have forgotten them, any more than they did the other stars they observed on this side of the line.[26]

This univocal polemic against the old authorities, their systematic destruction in the name of a sacrosanct autopsy, was not without its dangers. It amounted to a denial of any idea of progress, in favour of a revolution in knowledge that would necessarily proceed by way of the ubiquitist and totalitarian experience of the voyager-writer. With this virtually terrorist manifesto, situated right at the beginning of his 'Universal Cosmography', Thevet would end up opposing the naked eye, in a quite arbitrary way, to instruments such as the astrolabe and globe that were none the less indispensable to the pilot and practical mariner. His claims for a unique, 'naïve' experience thus bordered on arrant obscurantism.

The all-powerful gaze

The cosmographical is synonymous with the small scale, in the cartographical sense of the term. It presupposes that one can assume the ideal gaze of the Creator upon his world, or that one can transport oneself, like Menippus, into the lunar realms. In other words, there exist for the cosmographer two privileged means of access to the knowledge he seeks: that opened up by ecstasy, and that opened up by satire.

Let us briefly examine the latter, in the *Nouvelles des regions de la lune* ('News from the Regions of the Moon'): a sort of continuation of the *Satyre Ménippée* ('Menippean Satire') published in 1604, whose anti-Hispanic barbs borrow the style of Rabelais and the fictional theme of Lucian. Three legendary pilgrims, Aliboron, the Free Archer of Bagnolet and Roger Bon Temps, encounter the anonymous narrator 'on the highway that aims for Mirebeau', and in his company travel to the moon. From there they watch, through a trapdoor at their feet, the sinking of Spain's Invincible Armada. The *galaces* (heavy galleys) and gallions, carracks and *carraquillons*, and the brilliant flames resemble from afar butterflies fluttering about in torment, and the crewmen who hurl themselves into the water look like miniature eggs or droppings.[27] Just as Lucian, in his *Icaromenippus* and 'True History', mocked the geographers of his time, the anonymous follower who wrote the 'Menippean Satire' associates the 'lunatic' with the name of Thevet, 'who has seen invisible things'.[28] This process of reduction by distanciation was an adequate weapon with which to knock down pretensions to human grandeur, and in the event to ridicule the aspirations of 'his Satanic Majesty' Philip II to universal monarchy. Accordingly, the satirical emulator of the 'French Lucian'[29] parodies the rival ambitions of the hymn-singing poet and the cosmographer: both pretend to raise themselves into the heavenly empyrean and both privilege, in their representation of the cosmos, an all-embracing vision over particular details.

The phenomenon of cosmographical ascension could not be better described than by Agrippa d'Aubigné in his second book of *Tragiques*, where Virtue, addressing the poet in a dream, promises him an overarching view of the world and its history:

> I would make your spirit fly into the clouds
> That you might see the earth from the viewpoint of
> Scipio, when he was ravished by love of my name.[30]

At such an altitude, 'The world is but a pox, an atom of France.'[31]

Recognizable here is an echo of 'Scipio's Dream', that philosophical myth inserted into Cicero's *Republic*, and propagated during the Middle Ages and Renaissance through the commentary on it by Macrobius. Here, the soul's voyage into the highest sphere allows it to attain perfect lucidity; the limits of the universe and the ends of history are simultaneously revealed to it. The vanity of human torments, the useless agitation that envelops the terrestrial atom in a continuous whirlwind, become immediately perceptible to it, as do the rewards for the just and the punishment meted out to the wicked in the afterlife. Meanwhile, the harmonious music of the crystalline spheres revolving before the transported spectator plunges him into a lasting state of ravishment.

Such dream-fictions manifest the extraordinary operation that consists of elevating oneself above the universe in order to understand it and describe it in its totality. It is an operation that defies plausibility, and about which one tends to speak ironically; simple human forces cannot suffice. Thus, d'Aubigné attributes his flight through the air to Virtue or, elsewhere in the poem, to divine Providence. Only an active state of grace is able to transport the swooning soul, totally abandoned to God, back to its place of origin – that distant homeland from which terrestrial life has exiled it. Then, appearances are dissolved into a truth that escapes the view of ordinary mortals. And Admiral Coligny, who on earth was murdered, castrated and dragged through the mud on the morning of the St Bartholomew massacre, could in heaven enjoy such a derisory spectacle, one that only his killers, hunting down a ghost, could treat as a tragedy.[32] No doubt Thevet and the geographers of his age did not share the mystical bent of that Huguenot poet, who, on reaching the climax of his sevenfold epic, 'ecstatically swooned in the bosom of his God'.[33]

But what in d'Aubigné's case arose from a theology of history (as an eschatological progression towards a revolution that overthrows it, overturning its illusory appearances) corresponds, in the case of the savant, to a sort of a priori imperative. It is a transfer of scales that, from the outset, sets up the necessary frame of investigation. Without some voyage of the soul, there can be no instantaneous point of view over the cosmos. The kinship between cosmography and sacred poetry was therefore essential and primary.

In the beginning was the sphere, one might say; and it was for the members of these two professions, crowned with the laurels of royal recognition, to preside over its conquest. The circular or spherical figure of the cosmos, as has been noted in the case of Ronsard's *Hymnes*, was linked to 'no definite conception of the system of the world'.[34] From Aristotle it borrowed a progressive system of the four elements: from the immobile earth at the centre to fire at the periphery, via intermediary stages of water and air. It owed to Ptolemy geocentrism and the double motion, individual and general, of the celestial spheres; but it ignored his complex theory of epicycles and excentrics.

In fact, its extreme simplification of the cosmos pointed in the direction of an immediate intellection, an instantaneous possession. It facilitated the operation of miniaturization that Ronsard, in his 'Hymn to Philosophy', describes in a striking turn of phrase: thanks to Philosophy.

> The whole of heaven descends on to earth
> And its grandeur, grasped in a single sphere that
> (Great miracle) constrains so many stars,
> Like a plaything is placed in our hands.[35]

The stars are captives of the wooden sphere; just as, a little earlier in the same hymn, demons had been of the iron ring of enchanters. From the magic ring of the sorcerer to the celestial sphere of the cosmographer, the transition is ensured by round forms that express the plenitude of a power, the closure on to itself of a universal empire.

It is evident that Ronsard did not at all share the theological pessimism of d'Aubigné, any more than his thoroughly Calvinist caution on the question of man's capacities. The extent of the latter was not, in his view, so feeble that they could not, by magical means, force nature and imprison her within a restricted space – virtually containing, even domesticating her. His praise of philosophy – the general term that designated all branches of the study of nature, and consequently embraced the related disciplines of cosmography and epideictic poetry – was in the sense of a glorification of the human mind, as being capable of rivalling the Creator in his sovereignty over the world. To the traditional theme of the soul's ascension towards the cosmos, Ronsard lends a Promethean resonance:[36] now, by a movement of return, it is the cosmos that descends as a whole on to the philosopher's desk, for him to grasp it and reveal its secrets.

Rather than towards d'Aubigné, then – of whose poetic work, long unpublished, he was no doubt unaware – it was towards Ronsard that the Catholic Thevet inclined, both by temperament and community of religion, and in his profession as a cosmographer. Between Ronsard and Thevet the relationship was one of exchange of services and of dependence on the same protectors: in the first place the queen, Catherine de Médicis; but also powerful patrons like the cardinals of Lorraine and Bourbon, Chancellor Michel de l'Hospital, or the procurator-general Gilles Bourdin. Their social links and movement in the same circles gave rise to a community of intellectual concerns. The same symbolics is at work in Ronsard's *Hymnes* of 1555 as in Thevet's great *Cosmographie* twenty years later. Was not an inventory of the world, after all, an effective way of contributing to its celebration? A cosmographic encyclical was common to the designs of both hymnal poetry and the descriptive project of the universal geographer.

Thus we can understand the latter's use of prefatory quotations drawn from the principal members of the Pléiade and their followers, from Jean Dorat to Ronsard and from Jodelle to Guy Le Fèvre de la Boderie. These can be counted by the dozen at the opening of Thevet's works: two poems in 1554 in the *Cosmographie de Levant*, three in the *Singularitez* of 1557; but sixteen heading the *Cosmographie universelle*, and then seventeen for the *Vrais Pourtraits et Vies des hommes illustres* (1584). This poetic garlandry, celebrating the marriage of humanism and the new cosmos, exalted the twin figures of the visionary and the voyager. Following an image in favour among representatives of the generation that flourished with the reign of Henri II, Orpheus and Jason sailed off together in quest of other golden fleeces. A second Tiphys was their pilot; the latter thenceforward borrowed the face, voice and even name of that peripatetic Franciscan, the (almost homonymous) Thevet from Angoulême. Fulfilling the oracle contained in Virgil's Fourth Eclogue, Thevet – the 'other Tiphys' – was called to renew the world through his voyages. In fact, as Jean Dorat and Guy Le Fèvre de la Boderie remarked, his peregrinations inscribed within the circle of the sphere a cross: that of his itineraries first to the Orient, then to the south and the far west.[37]

The circular figure that cosmography privileged, by combining it with the cross that structures the world map, defined the perfection of a grasp as political as it was intellectual; 'For nothing is excellent in this world, that is not round.'[38]

The latter maxim of Ronsard, referring simultaneously to the global sphere, the king's crown and the priest's tonsure, postulates a sovereignty that would be more than geographical, but 'general and cosmic'.[39] Such 'cosmocracy', following a model borrowed from the thought of the ancients and realized from the outset of Augustus's reign, was an ideal realization of the poetic cosmography that was common to the poet of the 'Hymn to the Heavens' and to the indefatigable explorer of four continents, Thevet.

Now with the latter, the myth led into a political ambition; which is why the 'fictions of poets' with which he upheld his glory, and which his detractors attacked, should at all costs be taken seriously. Between Prince and Poet, 'each fascinated by the other' on account of their respective secrets,[40] was situated the Cosmographer, who performed a sort of instrumental transition between the two realms. Having access to the power of the Prince by the transmission (or, on the contrary, the retention) of strategic information, he was in a relationship of strict cousinage with the Poet. The unlimited power that a poet such as Ronsard, Dorat or Le Fèvre de la Boderie communicated to the king, as mythical sovereign of the terrestrial orb, found the beginnings of its realization in the map or sphere that was dedicated to the monarch, framed by his arms and traversed by his ships, and that opened up to his dreams of empire a space of intervention stretching to the limits of the terraqueous globe. The *mappa mundi* was a representation, at the same time hyperbolic and instantaneous, of an empire without limits; a concrete programme of military action, reckoning up places and deploying across an oriented space the dynamism of conquests to come. At the junction of an inspired poetics

and the most realist strategic calculus, cosmography embraced metaphor and hyperbole as active figures that made it possible, through
the efficacy of discourse and image, to transform the world.

Thus, too, we can understand that the excess of rhetoric presiding
over the geography of the early cosmographers was denounced by
the partisans of what one might call, with George Huppert, the 'new
history' arising in the late sixteenth century. From Jean de Léry, the
historian of French Brazil, to Lancelot Voisin de La Popelinière and
the Englishman Richard Hakluyt,[41] they all stigmatized a savant who
tried to usurp the glory of the Demiurge and based his authority
indifferently on the mathematician's instruments or on a treatise on
tropes. To accusations of this methodological order were indeed added,
from 1575, those of blasphemy. From the pen of François de
Belleforest, the author of a competing *Cosmographie* published in the
same year, and from that of Ludwig Camerarius writing on 22 July
to Hubert Languet, came the charge of impiety that would unite
French Catholics and German Protestants in unanimous condemnation
of the swollen-headed cosmographer.[42]

Now this was the effect of a risk knowingly taken by Thevet; the
inevitable consequence of his pretension to an exclusive reign over
things. By the bridge it threw between Poet and Prince, and between
the myth of universal sovereignty and its practical exercise, cosmography tended to translate into fact the ancient dreams of Alexandrian
or Roman cosmocracy. Thanks to the great discoveries, a suddenly
enlarged *oikoumene* coincided at last with the terrestrial sphere. Thevet
undertook to seize this historic chance and to make himself the
spokesman of a truly universal monarchy: that which fell by rights to
the France of the Valois kings. But, too pragmatic to share for long
that illusion, and too cunning to be the man of a single cause, he left
to flatterers of the moment the task of announcing that radiant future.
The poetic upwelling of the 1560s, published with a fifteen-year delay
in the *Cosmographie universelle*, reflected an old relation of forces
and proclaimed, at the time of the wars of religion, an optimism that
had already become anachronistic.

From then on, the cosmographer's design went against a movement
that had begun during his lifetime – and in the midst of his own social
milieu – in the development of the natural sciences. As François de
Dainville recalls, the Council of Trent had for its aim and its effect to
reconcile the domain of sensible experience with the principle of
authority. No doubt the 'situational epistemology' it imposed, of which
Galileo was the most famous victim, did not exactly contribute to the
growth of knowledge.[43] Control by ecclesiastical authority, wherever
it was established, imprisoned physics in the pillory of Aristotelianism

and confirmed the old attachment to Ptolemy's system, which the sacrilegious hypothesis of Copernicus had not succeeded in over-throwing. But even while obliging the savant to maintain an elementary prudence, those imperious boundaries did not forbid the investigation of natural facts. The return to intellectual discipline could to some extent encourage research, for example in the specialized domains of botany or astronomy. Thus a retreat was beaten from a general science, whose principles were no longer immediately in question, towards particular and practical fields of knowledge that were deeply dug over, to the extent that each allowed. Yet Thevet, unshakeable in his solitary pride and resolutely deaf to what Jesuit historiography calls, rather prudishly, the 'motions of Christian humanism',[44] would not for nothing renounce his universalist prerogatives.

Unlike the erudite libertines of his century, such as Michael Servetus or Giordano Bruno, who gave way to the temptations of speculative philosophy, the cosmographer to the kings of France barely moved away from the most tangible realities, being visibly ill at ease when an adversary forced him to fight on the plane of ideas. Always reluctant to speak in a theological, legal or even political idiom, he nevertheless committed blasphemy after blasphemy, emboldened by the experience that had taken him to the holy places. This even led him to contradict the Scriptures or the Church Fathers: at Bethlehem the manger was a cave, he said; and Jerusalem was not the centre of the world; nor was there any lion at the gates of Gaza nor whale in the Mediterranean, despite what the stories of Samson and Jonah affirmed.[45] He showed the same insolence in regard to St Augustine and Lactantius, whose writings denied the existence of the antipodes in a way that modern knowledge had refuted. Thevet's blasphemous attitude, which his friends, like the orthodox Sorbonne theologian Gabriel Du Préau, were the first to rebuke,[46] was not however manifested in a revolt against God and dogma; in his case it proceeded, at what seemed the most humble level, from the excessive role he accorded to that familiar servant of modern science: *experientia, rerum omnium magistra* (experience, the mistress of all).

The essence of Thevet's heterodoxy was, in the end, to have erected this personal experience into an absolute value, by whose measure he would be free to crush all authority, even the most venerable. By his repeated boasting Thevet called into question not only the teachings of Scripture; he compromised the very validity of all knowledge. One can agree, in this instance, with Father Dainville (whose analysis is not without apologetic intentions): the sort of regulation imposed on modern science from outside by the Church reborn at the Council of Trent would have exonerated the cosmographer of his 'perilous

wanderings'[47] – lateral wanderings, one might say, which, without
directly taking on theological dogma as such, sapped at its base the
whole pyramid of profane and sacred authorities.

The inexhaustible inquiry

Our problem, in truth, is in the first place one of method. Once we
fall back from poetic myth to the practical order of things, and from
Scipio's heaven to the geographer's study, we note that it is impossible to make the myopia of the human observer, whose visual field
is restricted, coincide with an enveloping intellectual vision of the
cosmos. These two extreme and opposing dimensions of geographical
investigation cannot both be encompassed within the same concrete
and immediate instance of experience. In other words, if topography,
the 'particular narration' so prized by Montaigne,[48] indeed arises from
the daily practice of individuals who travel, look at and take notes on
the ever-changing spectacle that surrounds them, then cosmography
supposes, by contrast, the mediation of a theoretical model and a
recognized scientific tradition – in other words, of analytical conventions. Faced with the landscape-object that it evaluates immediately,
topography stands alone; but to attain a total comprehension of his
object-world, the cosmographer needs auxiliaries and forerunners.
His profession places him at the end of a chain of workers on shared
tasks. As a product of the discipline's history, he also depends on
relays on the ground: a network of informers, whose observations he
attempts to recoup.

This apotheosis of the pilot on his ship, cross-staff in hand, appeared
as the emblem of a wilful oscillation between the most divergent
orders of magnitude. Thevet's vision was a deliberately floating one,
that tended to efface thresholds – the planes of rupture between one
scale and another.

Analogous to the posture of a pilot on his caravel was that of the
'insulist' on his island (a space hardly more stable than the bridge of
a ship); one that authorized the same transgression of scales. Climb-
ing to the top of the highest eminence, such an individual could see
the coast deployed symmetrically around him, inscribed on the
sprawling sea as if on the rectitude of a map. So it was with Thevet's
ascension of the 'rather high mountain' perched 'like a pyramid' on
the 'Isle of Rats' at four degrees north latitude. It allowed our intrepid
cosmographer to accede, beyond a merely topographical vision of the
immediate space (an islet infested with rodents), to a contemplation
of the cosmos, grasped in all its workings and its multiplicity. By dint

of a 'great labour' and of 'incommodities' surmounted at the risk of his life, he triumphed over the place and its attempt to ensnare him. During two nights spent under the stars, he enjoyed the spectacle of a universe laid out for his gaze. From this 'Angoumoisine Mount' (named after his home town – fittingly, in view of the symbolic conquest he thus effected), he penetrated with his naked eye the secrets of cosmographical science that academic geographers had searched for in vain.

His form of emphasis marks, here again, a passage from one order to another, from the local scale to a global scale:

> It was from this mountain that I saw the stars which lie close to the Arctic circle – that is, none other than the true revolution of that pivot of the ecliptic or Zodiac that is as distant from the Arctic Pole as the Tropic of Cancer is from the Equinoctial [equator], namely, twenty-three degrees and a half. So much can I say, in addition, of the Antarctic circle that bears the same name.[49]

In the present case, the anecdote (apparently suppositious, since it refers to a hypothetical 'first voyage' in 1550)[50] is intended to render plausible an abrupt change of point of view, expanding the field of vision from local experience to a universality of knowledge. But most often, Thevet barely escapes getting tangled up in such narrative precautions, resulting in the peculiar telescopic condensations in which he pretends to bear ocular witness to things that actually have nothing whatever to do with the visual experience of a traveller.

Say, for example, it was a matter of proving that the ocean is surrounded by lands and not vice versa, as the ancients had believed. Here it sufficed for Thevet to refer to his travels in the four parts of the world in order to dispense, in his view, with twenty centuries of cosmographic tradition. It was through the 'essays' – in Montaigne's sense, of an experiment or test – of his long-distance navigations that Thevet could assert that the Ocean was no such aquatic girdle bounding the *oikoumene* but, on the contrary, a second and larger Mediterranean, linking together the peoples scattered around its shores: 'Yet through my navigations *I have essayed*, not only that there was land, but as well that the sea was so bounded by it that there was no further water to be seen, as at the coast of Antarctica.'[51]

By dint of a spatial and temporal ellipsis, the fleeting moment of the voyage opens on to a fully abstract mode of evidence, ordinarily impossible to comprehend without the intermediary of a *mappa mundi* and mathematical instruments, and without the necessary withdrawal into the labours of the study: 'Now that I have found lands of such great extent, why should I continue to say that it is Ocean that

surrounds the earth? Since, on the contrary, *I have seen with my eyes*
the Ocean performing a sort of quick turn and turning back on itself
from west to east.'[52]

This confusion of scales meant that the pilot's gaze, scanning from
a great distance the curve of a gulf or the inflection of a line of broken
water, suddenly coincided with that of the cartographer in his study,
contemplating on a sphere the great masses of emerged lands that
were simplified by their miniaturized representation on the globe.

Indeed it is difficult to see how, without having recourse to the
ubiquitist vision of a Demiurge with his 'hands so great that in one
he could hold the whole world, and with two or three fingers turn the
entire earth',[53] the practical navigator, with his eye fixed on the horizon or raised to the heavens, could deliver coherent, uniform and
general information about the terraqueous globe. That meant having
to look beyond minor incidents of the voyage, his subjectivity as
an observer and, more generally, the limited capacities of the sense
organs.

But Thevet needed that fiction, from a standpoint at once particular
and general, personal and totally objective, in order to topple the
pretensions of ancient geography. The triumphant and omniscient
autopsy of a pilot 'in the seat of his ship', or of the insulist at the
summit of his island, served to condemn the reveries of the ancients
as well as the speculations of his vexatious contemporaries. Symptomatic of this need is his development on the way the earth encloses
the sea, at the beginning of the *Cosmographie universelle*, which
follows a critique, in the most traditional Renaissance manner, of
vana gloria. The double example of Aristotle and Empedocles killing
themselves through pride and a mad love of glory (the former by
throwing himself into the Aegean Sea, the latter by plunging into the
crater of Mount Etna) denounces, following a well-proven topos of
Christian humanism,[54] the vanity of a science whose encyclopaedic
vocation rendered it incapable of fixing human limits for itself. Not
noticing that this accusation of hubris could be turned against himself
(as it was after 1575, by Belleforest and Ludwig Camerarius), Thevet
stigmatized those who vainly tried to penetrate the secrets of nature,
such as the origin of the 'currents' and tides of the sea, instead of
commending themselves on such delicate questions to the infinite
prudence of God.

However, a first difference emerges here between Thevet and the
philosophers of antiquity: it is that he renounces their ambition of
plumbing the occult depths of the world, in order to concentrate on
mastering its surface by elevating his gaze to the heavens. His tendency to excess led him to follow Icarus in his flight over the sea,

rather than the shadowy traces of Aristotle or Empedocles who, engulfed as they were by the earth, became sealed up in what might appropriately be called an 'abyssal knowledge'.[55]

The cosmographer's other major difference from the ancients was that his rashness seemed to be justified, in a certain way, by Revelation. The Scriptures taught man that the world belonged to him in its totality, without exceptions of creature or territory. To recapture for man's gaze this universal homeland of the earth was the sacred task to which the cosmographer's labours were devoted. In his words, indeed, 'after theology there is no science that could have a greater virtue of making us understand the divine grandeur and power, and of causing us to admire it, than this one.'[56] It was a matter, thenceforward, of accentuating his rupture with a philosophy closed on to itself and which, having shown itself incapable of extending the field of investigation within the limits prescribed by Adam, denied that the world was knowable from one pole to the other.

Thevet returns time and again to this theme, and his *Singularitez* devote a whole chapter to it. It was through their blindness and presumption that 'natural philosophers', from 'Thales, Pythagoras and Aristotle' to Pliny and Ptolemy, had affirmed without proof that the torrid zone and the glacial zones – that is to say, three fifths of the earth's surface – were uninhabitable.[57] Scripture (which Thevet elsewhere has no scruples about denying as an authority) and the experience of the moderns proved the contrary; they were here united in the same movement of global enlargement and eradication of a past of error and superstitious restriction. In this context, his illustration from the Genesis story and his praise of navigation naturally formed a piece:

> Also man was created of God for that he might dwell and lyve in what parte of the worlde he woulde, were it hote, colde or temperate: for he him selfe sayde to our first parentes: Growe, increase and multiplie. The experience furthermore teacheth, (as many tymes we have sayde) howe large the worlde is, and commendable to all creatures, the which we may see by the continuall Navigations on the sea, and by the long journeys on the lande.[58]

The ancients had combined two seemingly contradictory wrongs. To the arrogance of a vain and indiscreet search for causes, they added a lack of curiosity about distant spaces. In this regard Thevet would have agreed with Montaigne when he ironizes about 'idle gossipers' and prefers to the irrational pursuit of causes an inquiry into the inexhaustible variety of things.[59] The project of cosmography would, in the end, be to restore to human knowledge that horizontal

dimension of which the pedantry of the ancients had largely deprived
it.

To achieve this end, it could do no better than radically overturn
all their rash affirmations: for example, against the opinion of Aristotle and in conformity with Avicenna and Albertus Magnus, Thevet
concludes that the torrid zone is not only habitable, but 'better, more
commodious, and more wholesome for our humayne life, than any
others'.[60] The same taste for polemical paradox leads him to turn
inside out the appearance of the world map, inverting the respective
spaces the ancients had assigned to water and emerged lands. From a
central island, the earth is inverted into a belt that encloses everywhere
an ocean reduced to the contours of a second Mediterranean or, better,
of a vast Caspian Sea.[61]

This spirit of contradiction, enlivened by the principle of autopsy,
obliged Thevet ceaselessly to straddle the great gulf between *quantitas*
and *qualitas*. By way of rhetorical acrobatics he tried to unite in his
hand the sphere as it appeared to the Creator who moved it, and the
narrow compass of human experience. As we know, he distrusted
abstraction. That is why he preferred, to universal maps in plane
projection, the 'globe and round sphere' that was a miniaturized
analogue of the world, capable of rendering to the cosmographer the
aspect, consistency and volume of the earth as it would appear to
God's gaze. From the globe that he could turn with his fingers like
the Demiurge, he drew the 'assured judgement' that he opposed, for
example, to those who pretended to designate America by the
problematical term 'Indies'.[62]

It would seem that our cosmographer ignored, in this, the correct
distance that, according to Pascal, a savant should (like a gentleman)
maintain between himself and any object.[63] Judged by the standards
of the *grand siècle*, his fault can be considered one of scientific
method, no doubt; but even more, one of an aesthetic, and hence a
social, order. For the lover of pictures, as for the physician, 'too much
distance' and 'too much proximity' equally hinder an adequate vision.[64]
Yet these were the mutually untenable positions required by the blind
ambition of Thevet's universal science of the concrete. Alternately
affected by myopia and long-sightedness, the cosmographer sometimes
had his eye fixed on the singular and unique details with which the
universe swarms, and at other times tried to embrace the whole in an
instant. Little aware of his limitations, and incapable of holding himself
to the 'middle ground between two extremes' that defined, for the
author of the *Pensées*, the right balance of the human mind,[65] he
pretended simultaneously to occupy both ends of the field.

Thus it was that when his autopsy tumbled from a general consideration of the world to the localized spectacle of a particular singularity, it tended to lose itself in sterile repetition. In trying to be everywhere at once, Thevet never managed to prove his point by the redundancies of his fictive experience. His doubling-up of the voyage to Brazil is symptomatic of this; as are his two crossings of the Sinai desert in the direction of Mecca,[66] or the naming of two 'Thevet Islands' opposite Canada and Brazil;[67] not forgetting the Angoumoisine Mount mentioned earlier, which came into being one fine night in 1550 a few degrees above the equator. In the duplication characteristic of this experimental backing one recognizes the distinctive traits of his imaginary transports to the ends of the earth.

Also symptomatic is the haste with which he settles on the spot the most diverse questions of natural philosophy. Thus, on the highly controversial problem of the sources of the Nile, about which Herodotus, Aristotle, Pliny and many others had erred in a fine chorus and with the most contradictory reasons, Thevet answers on behalf of the reader of such a cacophony by declaring haughtily, 'You may ask what Thevet concludes from all this. Nothing, other than what I know and have observed in the two years and nine months and more, during which I lived and philosophized in Egypt.'[68] (Here the former Franciscan's stay in Egypt, which barely extended beyond the winter of 1551-2, is drawn out to the unlikely dimension of almost three years.)

Again, in planning to check Pliny's chapter on the ostrich and encountering in Cairo one of these 'savage birds', he hastens to put the creature's proverbial gluttony to the test by presenting it with a piece of iron, 'which it found so good that it choked on it'.[69]

Hence the intrepid traveller's indignation in the face of the crazy ideas of the learned, who 'mock me – the one who knows the opposite to be true, because I have experienced it'. Indeed, the experimental zeal of Thevet unleashed a veritable slaughter. To refute Dioscorides and his commentator Mattioli on the subject of the density of the waters of the Dead Sea, which were said not to engulf a body cast into them, he pours into them a whole series of diverse objects, from the 'bones and heads of horses and camels, numbering over a thousand' to 'three of our people' massacred by Arab bandits; passing by way of the pack donkey of a Nestorian Christian and the 'Turkish' boots of one of his Christian companions.[70] Animated by a sort of rage to prove, Thevet strings together anecdote after anecdote: the ambush of the porters who served as guides, a dispute between some drunken pilgrims squabbling over a 'bottleful of wine', or the surprise attack by some desert warriors. He multiplies to 'five times' his visits to the 'lake' mentioned, and metamorphoses it into a bottomless repository: 'All these things', he concludes, 'did not fail to sink immediately to the bottom, and were lost to view.' Struck with all these confounding proofs, the unfortunate 'Matthiole' would surely not hesitate to lay down his arms.

The qualitative regression one observes here engendered a proliferation of the topographic viewpoint, to the detriment of the overall vision. From the rigour of a cosmographical tract one falls back, it seems, into storytelling. But beyond the doubtless involuntary nature of this reversion, Thevet's savoury anecdote betrays his attachment to what was at the same time an aesthetics and a principle of description of the world: an admirable, sparkling *varietas*. To it would belong his task of setting, like gems, the regular and all too uniform spaces bounded by the canvas of the sphere.

Cosmetics and cosmography

An ancient etymological play on the Greek word *kosmos*, 'which also means an Ornament or, if you will, something beautiful, pleasant and delectable',[71] tended to assimilate the universe to a visual spectacle, precious in its inexhaustible diversity and reserved by God from the beginning for man's pleasure and instruction.[72] This pedagogical scheme of the Creator relied on the image – or, more precisely, on the innumerable and varied images deployed, from base to summit, by the immense scale of beings – in order to make the human mind understand, by successive inductions, the unity of the great Whole. It was in turn imitated by writers of 'natural philosophy', naturalists, doctors and geographers. Fortified by their certainty of the principle that cosmetics and cosmography proceed from the same divine principle of variety, they all made of their learned descriptions that 'loosely assembled marquetry' dear to Montaigne in his *Essais*, in which a scattering of singular objects forbade any too precise taxonomy.

To restore, within the apparent disorder of an exposition, the euphoric diversity of the 'theatre of universal nature'[73] – such was the avowed project of the compilers of 'singularities'. In modelling itself on the mannerist ornamentation, costumes or gardens that in the first place 'made up' the world, and in seeking to repeat through them the miracle of divine pedagogy, such literature quite naturally based itself on an aesthetics of variety and entanglement. It therefore had to mix classificatory codes and weave into a single fabric the threads of heterogeneous discourses. Such a programme was sketched out by Thevet from the dedicatory epistle of his *Cosmographie de Levant*, and taken up without noticeable changes in the preface to his *Cosmographie universelle* twenty years later.[74] This project, following the model of Solinus's *Polyhistor*, involved an alternation of subjects in such a way as to 'recreate human understanding', the enemy of uniformity – like the earth itself, which requires a 'mutation of seeds'.

Thus, a secondary diversity of the text was to be created out of the primary diversity of the world, and based on the model of the latter. The table of chapters of the *Cosmographie de Levant* offers a concrete illustration of this procedure: here are evoked in turn 'Istria and Esclavonia', with their powerful and voracious inhabitants (chapter 4), 'Wolves' (chapter 5), 'transformations' that can be interpreted as lycanthropy (chapter 6), 'Bears' (chapter 7), 'an assault and combat we had at sea with the Turks' (chapter 8), Candia (Crete), earthquakes, hares, Constantinople, lions, tigers, camels, elephants and the origin of tides. The fluid plot – subject to many departures and returns – of a trip from Venice to Jerusalem allowed him to pile up pell-mell encyclopaedic material borrowed from Pliny's *Natural History*, some etymological 'lessons' on the names of the hare (15), the Colossians (31), or Marseille (58); some adages 'illustrating' the Cretans (9), Mount Athos (14), the Spartan virtues (27), botanico-theological speculations on the banana-palm or Tree of Knowledge (51) and, finally, inserted here and there into this gallery of singular images, the scattered and always hypothetical traces of a personal experience.

Even if the matter of the *Singularitez de la France Antarctique* shows greater homogeneity, as a function of the novelty of the subject, the author still takes the liberty of weaving in portraits of animals – such as the lions of Mauritania (chapter 4), sea tortoises (14), or a fight between a rhinoceros and an elephant (22: fol. 41r) – into the itinerary of his voyage. The customs of foreign or legendary peoples – the idolatry of Africans (17: fol. 31r), the use of palm wine (11: fol. 20), a battle of Amazons (63: fol. 124v), or a victory of Canadians (82: fol. 164r) – interrupt and fragment the linear narrative of an expedition to the land of the Cannibals.

In these two works, Thevet has rejected a distinction that was in his mind earlier: that which opposes 'histories' to 'natural questions'.[75] The *Cosmographie de Levant* and the *Singularitez* resulted from a balancing of each against the other. It is evident that the pleasure of the text and the illusion of truth that he procures for the reader result from such an alternation. In mixing accounts of human adventures with the varied tableau of nature, our 'singularizer' avoids breaking the universal harmony with any caesura. The laws of history participate in those of physics and medicine: sciences to which he attaches a whole network of analogies. Hence the flecks of dynastic chronology inserted into the monumental inventory of a city, or the portraits of great men set into the topography of some islet in the Aegean Sea, amid the natural curiosities of the place.[76]

But in this direction, the project of the author of the *Singularitez* goes even further. Not only does he reveal himself to be in perfect accord with the sciences of his age, which perceived between divine, human and natural things a set of tangible correspondences allowing them to be ordered within the same nomenclature, but he sets about playing on this layering of diversity in order to reproduce, thereby, the divine play of nature.[77] The result that this produced inevitably makes one think of the Chinese encyclopaedia cited by Borges, which juxtaposes in the same place of knowledge the most incongruous rubrics, and thereby ruins all principles of classification.[78] This is how Thevet presents the list of engravings illustrating his set of oriental singularities: 'So you will see the figures and portraits of beasts, Pyramids, Ypodromes [Hippodromes], Colossi, Columns, and Obelisks, as near to the truth as it has been possible for me to provide.'[79]

No doubt the majority of objects recorded here belong, within the theoretically unlimited field of *rariora* (rarities), to the subcategory 'oriental antiquities'; but there are mixed into it from the outset 'beasts' that are none other than the elephants, giraffes or crocodiles that an early scientific tradition endowed with the most marvellous qualities. Natural history provides an escort for the monuments of human history. Thus, the Nile crocodile and the ichneumon (Pharaoh's rat) slip into the open gap between Pompey's column at Alexandria and the 'mummied body' discovered in a tomb at Saqqara (the illustration of which seems to be copied from a late Gothic tomb-figure).[80]

How can we fail to discern, furthermore, in the list of vestiges evoked, a chaotic juxtaposition of references to disparate places and historical monuments, from pharaonic Egypt to Alexandrian Rhodes, from Roman Alexandria to the Byzantine empire? Those complicated words like 'Ypodrome' or 'Obelisk', which designated around 1500 realities that were still fabulous for the average 'benevolent' and average reader, ultimately constitute purely verbal objects, whose sonorities are engendered ever more closely as their echoes resonate along the syntactic chain. In this exercise in Rabelaisian nomenclature, Colossus generates Column by, it seems, the repetition of a common radical; Ypodrome proceeds from pyramid by inverting the first two letters; and the Obelisk consummates, with its terminal erection, the alignment of Colossi with Columns by borrowing from them the pivotal vowel *o*. The onomastic play that represented, along with a vogue for anagrams and for the equivocal, one of the bases of the poetic science of the Renaissance,[81] here becomes the rule of a *dispositio*.

In placing such a poetic and linguistic principle at the command post of the production of scientific texts, Thevet discouraged in advance all construction of taxonomies. Or, more precisely, he multiplied to infinity their primary categories. In effect, even if it is not possible to give a meaning to the capital letters sported, in the list cited above, by each of the admirable words it contains, one can none the less recognize in them the substantives that designate so many singular, individual objects. Beyond the manifest rarity of the Ypodrome or the Obelisk (but elsewhere, people might say the same of the hare, the wolf or the crocodile) this presence of the singular erects each term into an unassimilable individuality.

Classification, unable thenceforward to handle anything but sets that supposed restricted generalizations, stopped only at the irreducible atomistic units constituted by 'singularities'. Short of this fragmentation, nothing was proper: neither classes nor intermediary sub-classes. In fact, each individual – animal, plant or artefact – defined its own class, and thereby revealed itself to be essentially different from an immediately contiguous class of objects. The collection of singularities resembles, in the end, an irregular and innumerable archipelago of individual categories: a string of islets that must one by one be identified, described, and filled with a primary name. Thus intended, cosmography logically ended up privileging the insular.[82] And so it would have been in the case of Thevet's work, if the patronage of the great had not abandoned him in his old age.

But the framing title of his work remains. That Thevet, from his first to his third work, retained the term 'cosmography' to designate such a methodical ordering of the particular, and that he believed from the start that he could apply it to the account of his voyage to the Levant, indicates the sort of ambition that animated his project: to plunge into the minute diversity of the world, to recognize its fundamental and inexhaustible heterogeneity, paradoxically joined up with the project of accounting for its totality. Placing himself from the outset under the sign of a two-faced *cosmos* – a universe and an ornament – amounted to marrying that global and instantaneous vision to an indefinite dispersion in the field of singular things. Even if he might lose himself among the latter, this encyclopaedist of the disparate (as Thevet in effect was) managed to recover in his project the 'delectable', shimmering unity of the Creation.

Having established these general principles of Thevet's poetics, we can now proceed to the practical elaboration of his work. How, in particular, did he resolve the fragmentation of a description stretched between the small and the grand scales, that embraced the world and set out to exhaust its inexhaustible substance?

A triple inquiry – philological, mythological and cartographical –

will allow an inventory to be established, in the course of a survey more systematic than chronological, of the different techniques used by the cosmographer and his assistants (researchers and stylists, scribes and engravers) in assembling the whole *oeuvre* and its various parts. In the end it will reveal – in its necessarily unfinished state – the 'scale model' of the cosmos towards which Thevet, throughout his life, advanced with the patient obstinacy of a worker.

2
Ancient Lessons:
A Bookish Orient

That there is no history but in your present memory, to which you will be
aided by the cosmography of those who have written of it.

Rabelais, Pantagruel

Thevet's fiction of erudition

Thevet's *Cosmography of the Levant,* the first work he published on
his return from the near East, is not strictly a voyage account.[1] Its
itinerary, a periplus from Venice to Marseille by way of Constantinople, Egypt and the Holy Land, serves as a framework in which to
set out, in relation to the geographical locations successively visited,
positions of a rhetorical, philosophical or moral order, borrowed from
compilations of the time.

Among the instruments used in the fabrication of this eminently
personal work, the *Lectiones antiquae* (readings, or lessons, from
antiquity) of Coelius Rhodiginus[2] occupy the front rank; at least if
one is permitted to decide among competing sources for the positions
especially favoured by the humanists. Given the frequency with which
they haunt Thevet's first *Cosmography*, as would be noted half a
century later by his German translator, Gregor Horst,[3] the 'Ancient
Lessons' merit study as an example of a method of work and writing.

At this early stage, Thevet's scavenging or *bricolage* bore neither on
native myths, as would be the case from the *Singularitez* ('New Found
Worlde') on; nor on the hyperbolic and disjointed elements of a
fantastic cartography, such as he would elaborate in the course of 300
charts of islands in the unfinished *Grand Insulaire*. It as yet played,
in a highly traditional (and uncharacteristic) manner, only on the
commonplaces of a 2,000-year-old mode of knowledge that had been
disseminated anew by compilers around the turn of the sixteenth
century.

First published in sixteen books at Venice in 1516, the 'Ancient
Lessons' of Coelius Rhodiginus took their definitive form in 1541
with an edition by Frobenius published posthumously at Basle.

Extending by now to over thirty books, these magistral lessons derived from the courses Lodovico Ricchieri of Rovigo (whence the
cognomen 'Rhodiginus') had given in his home town and in Bologna,
Ferrara, Vicenza, Padua and Milan between 1491 and 1525 (the year
of his death).[4] As digressive commentaries, often taking a word, adage
or juridical formula as their point of departure, his chapters generally
turned into an erudite reflection on some usage or philosophical
doctrine of the ancients, or an inquiry into natural history and geography, based on philological analysis. The study of *words* led to a
knowledge of *things*; and grammar was in this sense seen as a restorer
of the law, of medicine and even of theology; all disciplines clouded
by the passage of centuries.

In the thousands of *loci* that they examine and assemble into
concatenations of authorities, the 'Ancient Lessons' take their place
alongside the *Cornucopiae* of Niccolò Perotti (known as Sipontinus),
or the *Dictionarium* of Ambrogio Calepino, the famous 'Calepin'.[5]
An impressive fund of knowledge, 'such as a human life spent in
continuous study would hardly suffice to assemble',[6] the 'Ancient
Lessons' well deserved the subtitles 'Cornucopia' or 'Thesaurus' that
they are given in the 1542 edition. They were truly a horn of plenty
overflowing with the fruits of ancient Greek and Latin knowledge, an
inexhaustible treasure that generations of humanists, from Rabelais to
Montaigne, would draw on.[7] A hasty compiler like Pierre Boaistuau
of Nantes would go to it several times in putting together his 1558
Brief discours de l'excellence et dignité de l'homme;[8] while the
unclassifiable Étienne Tabourot, the lord of Accords, cites him in his
Bigarrures for a definition of the acrostic, along with the legendary
authority of Ennius and the Sibyls.[9] Somewhat nearer to Thevet and
to the cosmographical genre was Bartolomé de las Casas, the historian of the 'new Indies' and apostle of the Indians, who frequently
draws on Coelius Rhodiginus in his work as a chronicler and apolo-
gist.[10] An indispensable furnishing for any scholar's study, and found
in the libraries of numerous convents, the work of Coelius Rhodiginus
would be the source of a whole series of second-generation writings
as diverse as its own subject-matter. Not only did it provide a reference
work in which a missing citation or authority could be looked up, but
the less scrupulous compiler could take from it whole tracts of material he did not have the time or the patience to research elsewhere.
In the miniature library that they amounted to, he could find ready-
made the chains of authorities needed for any given topic. Associa-
tions between different topics were already woven together there, and
from the outset a hierarchy of texts was established; all the reader-
writer had to do was copy or continue it.

In thirty of the fifty-eight chapters of Thevet's *Cosmographie de*

Levant, its traces are clearly evident. Some of these chapters are even entirely based on it. Thus the short chapter XLVII entitled 'De la doctrine et exposicion des Hebrieux' (On the doctrines and theses of the Hebrews), which François Secret has recently studied, expounds on the Kabbala as interpreted by Pico della Mirandola; but does so through Rhodiginus's *lectio* X, 1 ('Hebraeorum doctrinam triplici calle incedere, tamultico, philosophico, cabalistico'), which Thevet merely takes over.[11]

Such cases, where the borrowing seems at the same time massive and continuous, are however rather rare. More often, Thevet takes advantage of the *Index vocum et rerum singularum cognitu necessarium* (index of names and subjects) included in every edition of the 'Lessons', to line up in the body of his chapter as many as a dozen separate references. Thus, on the subject of Sparta one seems to discern a reference to Plutarch and his Life of Lycurgus; yet our compiler has merely used, from the 'entries' opened to him by the said index under the headings Sparta and Lacedaemonium, nine topics derived from six different *lectiones: Hospitum expulsio a Sparta; Eurotas, Sparta, Lacedaemon, unde appellationem traxerint; Spartanorum mores explicantur; Spartanorum sacrificia praetenuia; Laconici breviloquii commendatio;* and *Pudicitiae commendatio.*[12]

The last-mentioned piece, in which the city of Laconia is not explicitly named, concerns a subject the scribe could also have found discussed in an earlier chapter about the island of Chios, whose women were famous throughout the ancient world for their chastity.[13] Recycling a reference in one of the two *lectiones* (XIII, 6 and 7) that had earlier served him on this topic of great moral import, Thevet (or at his behest, more likely, his friend and hack Belleforest) could thus have put aside a precious source of information for use later in his work. In this case, the use of summary notes allowed him to refine and vary his derivations from the *Lectiones antiquae*. The latter's index no doubt served as a point of departure, but its user shows that he was no slave to it.

Thevet's freely associative method – supported by the paratactic and arbitrary order of the alphabet, but modifying it as it went – has something in common with the recourse he had at the same time to the *Cornucopiae* of Perotti, the *Dictionary* of Calepino, or the *Thesauri* of Robert and Henri Estienne. For example, the first chapter of the *Cosmographie*, devoted to Mantua (though not at all a description of that city nor an account of a sojourn there by the author) combines, to the exclusion of all other sources (and, of course, of any personal remarks) the articles 'Mantua', 'Po' and 'Vesula' found in these various dictionaries that Thevet had at his disposal.[14]

There was a kinship, too, in the way the *Cosmographie de Levant* used commentaries on Pliny, on Solinus (by J. Camers or S. Münster) or on Pomponius Mela (by J. Vadianus). Ermolao Barbaro's *Castigationes* on Pliny and Pomponius Mela played the same role, which consisted of citing in the margin of the text the greatest possible number of authorities and maxims, without any real importance in themselves but simply forming the linear support for a parade of erudition that was infinitely extensible.[15]

Thevet allusively refers to the 'author of the antique lessons' in only one or two passages of the *Cosmographie de Levant*, but he uses him on almost every page and on the most varied topics: from Roman law to botany, from the proverbial fear of the hare to the interpretation of the Kabbala; from the bitterness of Corsican honey to the comparative study of Egyptian and Roman laws, or from the virtues of Nile water to a eulogy of the excellent government of Marseille.[16] In chapters that owe to Coelius Rhodiginus both their subject-matter and their structure, such as 'On Lacedaemonium' (XXVII) or 'On the Laws of the Egyptians' (XXXV), the latter is never mentioned. In the course of a parenthetic remark on the purifying qualities of the sea, which dissolves all crimes except that of parricide, Thevet furtively adds, 'as Catullus and Cicero attest'.[17] Yet this evocation of authorities summarizes an entire Lesson (XI, 22) of Rhodiginus, where the latter brings together a passage from Cicero's defence *Pro Roscio Amerino* (seeking, precisely, to 'cleanse' an innocent man of the accusation of parricide) and a fiery epigram of Catullus on the same subject.[18]

In Thevet's hands the interpolated clause 'as Catullus and Cicero attest' becomes incomprehensible; the reader without a solid humanist education would be incapable of finding on his own the references in question. But it becomes clear that the aim of these allusive (to say the least) references is not to promote scientific truth, or to resolve some profound aporia. They merely throw dust in the eyes of the 'benevolent reader' by a cheap display of erudition. The authority of a Catullus or a Cicero ceases to have any value as an authority. What do they 'attest' to, in fact? What is the point at issue, in a constantly digressive text that never leads to any summary or conclusion?

Such references have ornamental value: they correspond to the aesthetics of *varietas* proclaimed at the outset in the *Epistle to François de la Rochefoucauld*, and later recalled in the course of the work. In this way Thevet sought to densify his work of compilation; not only by piling up the most varied objects of study and the most diverse questions, but also by citing in a disorderly manner the greatest possible number of authors. All these proper names – much more numerous than the writers (at most, about thirty) actually consulted by the composer of the *Cosmographie* – fling open to the amazed reader the treasures of ancient wisdom. In this play of nomination and accumulation of authorities, assembled into irregular lists and not immediately verifiable, the 'author Thevet' was not the only beneficiary. For the counterpart to such prestige acquired on the cheap was the new dignity with which the reader was invested. The latter, a willing accomplice in an illusion that flattered him, could not help but enter into the 'fiction of erudition'. A pact of reading invited the public to acquiesce in a *trompe-l'oeil* that offered them a posy gathered at minimal cost, yet with inestimable symbolic value. It left them free to display in turn these rare and diverse flowers, to pride themselves on a knowledge twice stolen.

Exhibiting a borrowed humanism, Thevet's *Cosmographie de Levant* parasitized the robust compilations of European scholars of the turn of the sixteenth century. Fifty years on, it recycled the contents of that primary humanism for a public with no knowledge of Greek, nor perhaps even of Latin. Demanding no effort on the part of its potential readership, the *Cosmographie* adopted the recreational form of Solinus's *Polyhistor,* 'in order that a work composed of various subjects might better entertain the human mind, which, like the earth, needs a mutation of seed'.[19]

The density of authorities cited contributed to an effect of studied variegation. At the same time it procured, for cultural *parvenus* like Thevet, Belleforest and their readers, the social benefits of an encyclopaedic knowledge – one gained without effort and, as it were, by a game.

From emblem to cosmographical meditation

It was thus a question of ornamentation. Thevet's most summary structural procedure consisted of setting a precious stone stolen from the 'Ancient Lessons', without warning or transition, into the matter of his voyage to the Levant. 'I should not forget, here, the Jewish Doctors and Exegetes . . .': thus begins, *ex abrupto*, a digressive chapter on the Kabbala that delays, accordingly, the pilgrim's entry into Jerusalem.[20] This most brutal of transitions can be compared to another one, which opens a colourful page on the legendary Pygmies who rode goats and fought with cranes: 'I will write here a brief, but for all that no less true, chapter which you may find curious.'[21] It would appear that our author-compiler-interpreter was little concerned with providing smooth and insensible transitions from one topic to the next. Tightly enclosed in successive compartments, in a kind of geographical goose chase, the matters of inquiry are presented for what they are: curious objects brought from afar, and collected together for the pleasures of the eye and the mind.

Serious problems of classical philology inherited from the humanist of Rovigo are introduced with the same off-handedness: 'One suspects here something worthy of being understood; namely, whether the word Dasipus (which means a cony [rabbit]) is a general and appellative name, or a proper and special name.'[22] The question is an important one, and is inserted between the series of different *apelacions* of the hare in Greek or Latin and an evocation of the legendary habits of that animal.

The deictic adverb 'here' was admirably suited to this sort of local graft, by which an interpolated item was fixed into a certain point of the Levantine itinerary: 'Here, it is convenient to treat of the Colossus of Rhodes . . .' (chapter XXXI, 104, 20); 'Here, it is worth noting that the Egyptians were more careful than other nations about their burial' (chapter XLII, 155, 20–2); 'Here, I undertake (God willing) to give you a general and succinct description of Judaea . . .' (chapter XLV, 165, 26–7); 'Here, I shall in a few words describe Bethlehem, and Jericho' (chapter L, 180, 2–3).

More elaborate, but easily combined with this type, was the etymological or grammatical grafting that allowed Thevet to draw from the proper or common noun forming the title of a chapter (such as 'Mantua', 'hares', 'Athens', or 'pyramid') a 'truth' in the form of a history, an eponymous hero, a question of natural philosophy, or a philological controversy. The detail borrowed from the 'Ancient Lessons' would not only fill an empty compartment, it would also tap the reserves of meaning contained in any of the lexical stopovers on this pedagogical itinerary. Thus, Mantua calls up the memory of Manto, the daughter of Tiresias (I, 17, 2); hares give rise to a learned discussion of the origin and reasons for their name (XV, 53–4). Athens and the goddess Athena, or the pyramids and fire (which also diminishes as it rises), involve a reciprocal pertinence that enhances lateral resonances within the linear progression of the voyage (XXVIII, 91, 15 and XLI, 152, 12).

The process of emblematic or allegorical 'inlaying' corresponds to another order of motivation. It consists of linking to a toponym, or more generally to a name taken as a signifier, a moral signified. Its usage is rather frequent in the *Cosmographie de Levant*, which, by combining image and text, comes close at times to being a collection of emblems. Thus, 'portraits' of the lion, the elephant or the camel are the occasion for edifying developments on clemency, conjugal fidelity and sobriety.[23] As for the giraffe seen by the traveller at the castle of the Great Cairo, it is, as Angelo Poliziano taught, a 'simulacrum of the learned and the lettered'.[24] The association of the soul and body of the emblem – of *motto* and image – is then completed by a few lines of explanation: 'For they seem at first sight to be bitter, rude and troublesome; whereas by reason of the knowledge they possess, they are much more gracious, human and affable than other creatures with no knowledge of letters and virtues.'[25]

It will be apparent that such allegorical play prolongs that of etymological allusion. The giraffe had seemed to the Romans to be a mixture of the camel and the leopard (because of its dappled skin); whence its name of *camelopardalis*.[26] The emblematic significance that Thevet, after Angelo Poliziano, finds in the animal similarly rests on a dichotomy inscribed in its body, between its exaggerated forepart and more modest after part. The emblem thus proceeds from an analytical reading, element by element, of the anatomy of the 'monster'.

To back up these specific remarks establishing a relationship between the *Cosmographie de Levant* and the emblematic literature of the Renaissance, I should mention in addition an isolated borrowing the work makes from one of the theoreticians of the genre. In his preface, Thevet concludes his eulogy of vision – the most perfect of all the senses – with this somewhat obscure observation: 'In any case, we cannot deny that this so necessary sense shows us the differences between things, and that by it we arrive at a comparison of their quasi-living image with the half-dead letter [of their name].'[27]

Now this is merely an elliptical repetition of the preface Barthélemy Aneau had written for his *Décades* of emblematized animals, published in Lyon in 1549:

> And since (as King Candaulis says in Herodotus) the eyes give man more certain knowledge than his ears, I have portrayed and shaped as near as possible to their natural appearance, or to their proper description, the figures of these beasts, both tame and wild, familiar and foreign; so as to compare their quasi-living image with the half-dead letter [of their name] and to delight our corporal eyes with the sight of their picture, even as our spiritual understanding learns by reading.[28]

This contrasting encounter of the living image and the dead letter represented the very principle of emblematic syntax. In accordance with an analogy of Platonic inspiration, the image gave access to spiritual signification by way of the noblest of the five senses. From 'body' to 'soul' and from the pleasure of the eyes to the contentment of the soul, this complementary pedagogy of the graven image and the printed text allowed a two-stage initiation that joined the useful to the agreeable. As in the moralized bestiary that Barthélemy Aneau inherited from the Middle Ages and renewed, contaminating the (at the time) new genre of emblematics,[29] the *Cosmographie de Levant* repeatedly conjugates 'verbal knowledge' with 'imaginary knowledge' in such a way as to draw from natural philosophy its highest meaning.[30]

The emblematic function was not limited, for Thevet, to an inventory of the Great Turk's menagerie in Constantinople (chapters XVIII to XXII) or in Cairo (chapters XXXVIII, XXXIX) – or to theories about particular creatures that recalled (without strictly imitating) the zoological *Décades* of Aneau. It extended to the whole of the book by a continuous counterpoint of image and text, and by that procedure of incrustation that characterized the *emblema* in the original sense of the term.[31]

His emblematic borrowings from Coelius Rhodiginus bring a moral sense to the natural or cultural curiosities encysted in the course of his oriental periplus. The same might be said of the *Lesbia regula* or 'rule of Metelin' that was to be found in the *Adagia* of Erasmus,[32] which introduced into the contours of that Aegean island a politico-philosophical reflection on malleable laws (that conformed to local manners rather than rectifying them, as they should).[33]

The 'Egyptian judge' or *Aegypticus iudex*, meaning 'a good judge, who is not at all corrupted by gifts and presents',[34] takes his place in the same way, in one of the chapters our Franciscan devotes to his stay in Alexandria. Associated with another proverbial formula, also drawn from the 'Ancient Lessons', 'Virgo justicia est, prognata Jove' (Justice is a maiden, the child of Jupiter),[35] the emblem here allows the exposition to rise above a description of countries to the level of paraenesis, in a tendency that Thevet's scribe Belleforest would later develop in his own work as an author.

Erasmus's *Adagia*, which Thevet used alongside the uncountable treasures of the 'Ancient Lessons', perhaps bear even clearer witness to the emblematic function of such a marshalling of resources. The adage 'Aetna, Athon', which the humanist of Rotterdam drew from the *Noctes Atticae* of Aulus Gellius to signify an extreme world-weariness,[36] was quite naturally integrated into Thevet's chapter devoted to 'the Sacred Mountain, otherwise known as Athos'.[37] The case is interesting in so far as Thevet, unlike an Erasmus shut away in his study, was basing himself on the concrete reality of a peregrination. Now to decorate this geographical reality he used a text with allegorical and moral value, at the same time inverting the relation of priority between the comparing element and the compared. It was no longer geography that illustrated or symbolized a truth of an ethical order, but the latter sort of truth that was made to enrich and even define the place or geographical form encountered. In this, Thevet was inverting the customary use of the *Adagia*. Instead of seeking in them a moral lesson illustrated by an example, he departed from an illustrative example, a 'local signifier', inserting it in the appropriate place within his itinerary and adding the lesson that went with it. Wherever the path of description led, the geographical topos thus drew to itself, from one moment to the next, a rhetorical or moral topos.

Such an example, which is not an isolated one in the *Cosmographie*, thenceforward belonged to the genre of 'cosmographical meditation', to use the eminently suitable term of Gerard Mercator;[38] a genre represented by a whole lineage of geographer-theologians like Sebastian Münster, Joachim Vadianus, Josiah Simler and Jodocus Hondius. Such cosmographical meditation (which Thevet, for his part, would abandon early in his career) made the *mappa mundi* enter into the composition of pictural 'Vanities', where it featured alongside the cranium, bones, candle and hour-glass. It enabled one to read in the map – for example, by considering the unequal distribution of oceans and emerged lands – the admirable design of divine Providence.[39]

Thevet's privileged support and point of departure were, in this, the *Epitome trium terrae partium* by the Swiss Joachim Vadianus, the reformer of St Gallen, whose religious message makes its presence felt in large tracts of the *Cosmographie de Levant*.[40] It might be a question, for example, of depicting the spiritual ceremonies of the Libyan Pentapolis, or of the Egyptian Thebaid: here, Vadianus provided not only the catalogues of holy persons and ecclesiastical writers that Thevet needed, but furthermore gave, step by step, the key to a geographical reading of biblical history. In the course of the voyage, a local anamnesis favoured the revelation, beneath the pilgrim's feet as it were, of the evangelical life and the blessed times of the primitive Church. Cosmographical meditation was oriented into a somewhat different path by the moralization of a landscape or fertile countryside, a walled city, a stony desert or a luxuriant valley. Thus, the land of Gaza brings forth good wines and 'a great abundance of almonds and pomegranates'. This actual, very material fecundity brings forth in turn another one, of a spiritual order, 'by the seed of the divine word and the rain of the Apostolic doctrine'.[41] We find the same elevation from the temporal to the spiritual, from geographical space to its highest meaning, in the case of Jericho: 'This city was very rich and abounded in balm; but it was richer still in the odour of the divine word.'[42] Or again, of Antioch, the city of red lilies: 'But in truth Antioch has no rival more famous or better known in the production of so many beautiful flowers – of, that is to say, martyrs and confessors . . .'[43]

In the case of Gaza the similitude is accompanied by a chronological opposition, between a radiant past and a deceptive present. An identical contrast could be established between the topographical appearance of a place and its spiritual essence. So it was with the Arabian desert: a sterile and sandy place, but one infinitely richer than the gold and perfumes of the Arabia reputed to be 'Fortunate' when it came down to historical and moral meanings. For it was in that precious desert that the Hebrews had stayed for forty years after leaving Egypt, and in which was concluded the first Alliance: 'I dare say that this part should be preferred to Arabia Felix, for this reason alone: that it received the children of Israel after they had passed through the Red Sea with dry feet.'[44]

Proven in this way by the evangelical message, geography was turned inside out. Whether 'Deserted' or 'Fortunate', the two Arabias exchanged their respective qualities and their sites. The Christian topography of Vadianus reversed physical appearances in the interests of his pupil-reader's spiritual edification. Integrated into the marquetry of Thevet's voyage to the Levant, Vadianus's cosmographical meditations join in composition with 'common proverbs' garnered from Erasmus and with the lessons drawn from Coelius Rhodiginus. They thus melt into the general emblematics mapped out, at the crossroads of image and text, by the *Cosmography de Levant*.

Dialogisms

The 'author of the "Ancient Lessons"'[45] is, furthermore, often solicited for the last type of structural montage that needs to be examined here. This is the dialogism, a procedure by which the compiler opposes to an authority such as Pliny or Aristotle his opponents, both ancient and modern. From the 'Ancient Lessons' were drawn certain problems worthy of meditation: did elephants have joints in their legs? Should their tusks be regarded as teeth, or rather as horns? On the former question, Thevet did not hesitate to oppose the opinions of Strabo and St Ambrose in the name of experience, by quoting a famous anatomy lesson given a few years earlier at Damascus by the naturalist Pierre Gilles.[46] As for the second question, he inclined – with hardly any more originality – towards the view of Coelius Rhodiginus, who had decided in favour of horns.[47]

The most prolix illustration of such a structure is given in chapter XXXI of the *Cosmographie de Levant*, entitled 'On the Colossi of Rhodes', where nearly three lengthy pages are devoted to debating the thorny question of whether the Colossians addressed by the apostle Paul were the inhabitants of the town with the celebrated colossus, or those of Colossae in Phrygia. The outcome is known in advance, and of little moment. Raffaello Maffei (known as Volaterranus), on whom Thevet drew for all the prestigious names with which he decks out this fictitious dispute (Herodotus, Strabo, Pliny and St Jerome),[48] had already solved the problem in a most satisfactory manner: 'Colossis urbs apud Lycum fluvium, cujus meminit Herodotus . . . Hujus urbis meminit item noster Hieronymus. Ad hos igitur epistolae sunt Pauli, non ad Rhodios (ut vulgus putat).'[49]

All is said here in a few words; but what mattered for Thevet were the possibilities for amplification by dialogism that were contained in this slender correction. By going back to some of the authorities cited by Volaterranus – essentially, St Jerome on the Epistle to the Colossians – and naming them all (even if, in three cases out of four, they had nothing to do with the controversy in question), the compiler managed not only to furnish, but even to animate the subject of his *Cosmography*. The 'fiction of erudition' I mentioned above was here procured by the staging of a profuse and purely ostentatious dialectic.

Of more modest dimensions were his debates on the bears of Africa (chapter VII, p. 30) or on lycanthropy (chapter VI, p. 27); but they combined in the same manner to give the exposition colour and relief: 'One might ask, how do the Ethnics [heathen] bear on all of this, who feigned so many transformations? . . . I would reply that some can be found who, having a vitiated and corrupted imagination, actually believe that they have the form of a wolf . . .'

By turning to his own ends a doubt expressed by Pliny, Thevet hardly ran any risks in dealing with a matter that was among the most delicate of his time.

The repetition of these feigned disputes from chapter to chapter could, in the end, have become tedious for the reader. Thevet perceived the danger; and that is why he tried on at least one occasion to put the debate into context. To overcome the monotony of a journey on the Black Sea, 'a question that is worth relating was proposed to M. Jean Chaneau, a native of Poitiers, by a person in the company who was greatly enamoured of virtue: namely, whether fish can breathe or not.'[50]

It need hardly be added that this slender narrative pretext is effaced as soon as it is mentioned, leaving the field free for the 'common viewpoint' (in the event, that of Pliny, corroborated by a reading of Coelius Rhodiginus) and an exposition of it that fills most of the rest of the chapter.

Thevet almost admits to his ploy when, at the end of a philological discussion bearing on the different names of the hare, he concludes or avoids concluding:

Here you have, gentle Reader, authorities for one side and authorities for
the other; of which I have tried to inform you and show you the grounds
on which they base themselves. However, I wonder whether I should not
add that men easily contradict each other, often without any just or reasonable cause.[51]

It was unlikely that the benevolent reader would presume to decide
anything in place of an author who so insistently proclaimed his
modesty – notably, by declaring that 'in this affair, it is not for me
to judge such great personages.' But his purpose, it is evident, lay
elsewhere. The benefits of a dispute set out and propped up with
authorities taken (here again) from Coelius Rhodiginus, were not
of the sort to be translated into scientific advancement. Later, in the
Cosmographie universelle, Thevet would regard himself as the equal
of these authors and would not hesitate to abuse them. But for the
moment he affected to revere them, though without hesitating to
ascribe their vain desire for debate to the weakness of human nature.
Paraenesis, which has the last word in these leporine pages, was already for Thevet a means of raising himself to the stature – or even
slightly above it – of the 'great persons' that the hasty compiler knew
only by name.

Gregor Horst's Thevet

From the radiant dawn of the Renaissance to its vulgarizing apogee
in France, the trail leading from Erasmus's *Adagia* or the 'Ancient
Lessons' to Thevet's *Cosmographie de Levant* was marked by a re-
distribution of primary knowledge. From the voluminous erudite
miscellanies put together between 1490 and 1520 down to the pleasant
illustrated compendium of Belleforest–Thevet, the format and presen-
tation of the book changed, and the reading public was transformed
and enlarged. But the process did not stop there. Thevet, who was to
Erasmus and Coelius Rhodiginus something like what Solinus had
been to Pliny – an abbreviator endowed with a sense of the marvellous
– would himself meet with a rather similar fate.

   The immense popularity of works 'aping Pliny' from Carolingian
to Renaissance times is attested to by manuscripts and editions in
their dozens,[52] and had a late and fecund flowering at the beginning
of the sixteenth century. It was then that the rich commentaries
of Johannis Camerarius and Sebastian Münster rendered down, so
to speak, on to the meagre canvas of the *Polyhistor* – sometimes reducing it to a nomenclature of toponyms, as in the chapter on the
Aegean islands – the ample flesh of Aristotle or Pliny. The zeal of the

humanists completed the cycle. The summary was augmented by all its 'left-overs', so that it began to comprise a sum of information comparable to that contained in the original text!

One might think that the outcome of such an operation was nil. In fact, however, if science had progressed little from a strictly quantitative point of view, it henceforward obeyed a different principle of organization. The new disposition, multiplying grafts and offshoots, engendered a new practice of reading. The repeated incursion of proliferating marginalia into a text, affecting it in every part and encrusting themselves there in the form of reminders, notes and references, effectively blocked any serious reading. The dialectic that Thevet artificially introduced into the body of his compilation by way of rhetorical questions represented, there, the real exercise of reading. As a necessarily formal criticism, the latter tended to put into question the very notion of authority. This process culminated in the severe *castigationes* with which Ermolao Barbaro 'chastised' Pliny and Pomponius Mela. Later François de Belleforest, too, situating himself in the anti-authoritarian tradition that critical humanism inaugurated after the turn of the century, would attack with his distasteful glosses the 'Universal Cosmography' of the illustrious Sebastian Münster.[53]

Following Solinus's summary of Pliny, the bundle of ancient lessons, adages and other gems of erudition assembled by Thevet found its own commentator: Gregor Horst, the German humanist and physician to the Prince Elector of Hesse. Around 1613 he set out to fill the leisure hours forced on him by the diet of Ratisbon by translating into his native tongue the book Thevet had published in Lyon some sixty years earlier. The fruit of his labour was a *Cosmographia Orientis, das ist Beschreibung desz gantzen Morgenlandes*, published in 1617 at Giessen by Caspar Chemlin.[54] In this German version, the *Cosmographie de Levant* was enlarged by substantial marginalia, augmented with additional plates based on illustrations in Thevet's 1556 edition, and ornamented with a woodcut taken from a pamphlet by Joachim Strüppe, showing the 'true portrait of an Egyptian mummy'.[55]

Thevet's discreet conformity to the Reformation, for example in relating an episode like a storm at sea, was not displeasing to this Lutheran, who praised the 'piety' of an author who described the 'conjurations' of the common people as 'a mockery, and a thing more Ethnic than Christian': 'Allhie beweiset der Autor recht als ein frommer Christ, dass in allen unsern nöthen wir keinen anderen Helfer und Mitler suchen sollen, als allein Gott dan Allmächtigen.'[56]

With regard to the chapter on Antioch, Gregor Horst notes with satisfaction that Thevet does not agree with the official title of the

Jesuits, which to him seems to usurp a name reserved for the Son of God.[57] But elsewhere the translator dissociates himself from the evangelical Franciscan, when the latter summarizes the 'errors' of the patriarchs of Greece. For in his eyes, the Orthodox Greeks were less distant from the truth of the Gospel than the Catholics attached to the pope in Rome: 'Dass die Griechen in vielen der Evangelischen Warheit neher kommen als die dem Bapst zu Rom anhangen, ist allhie offenbar, wiewol der Autor es vor irrthumb helt.'[58]

The almost total ideological agreement between the early Thevet and the Hessian doctor, who freely cites Luther's Bible in order to correct the errors of the Vulgate (on the subject of the Pygmies, for example; of whom Ezekiel, *pace* St Jerome, never spoke)[59] no doubt contributed greatly to this unexpected encounter and revival after an interval of fifty years.

But the essential point lay elsewhere. Thevet's work recommended itself, as the translator assures us, by its compendious variety. Among the books on pilgrimages to the Holy Sepulchre, few were to be found that gathered together, so briefly and in so few words, so many memorable things.[60] What interested Gregor Horst were the numerous possibilities for commentary that were opened up by both this diversity of subjects and this concision of purpose. The *Cosmographie de Levant* was still preferable to the more voluminous *Observations* of Pierre Belon or the *Navigations et pérégrinations* of Nicolas de Nicolay, on account of the perfectly topical and unoriginal nature of its content. In his dedicatory epistle, the translator-commentator insisted in particular on the 'feine lustige *quaestiones*' (fine, amusing problems) that spangled the Frenchman's *Cosmographie*, and whose resolution was of the highest utility.[61] Now these erudite 'questions', whose quality and necessity were noted above, would allow the commentator in turn to parade the most diverse knowledge. Sometimes it would be a case of establishing connections with the eye-witness accounts of German travellers like Hans Jacob Breuning or Rudolph Kircher;[62] at other times, of pointing out historical information that recent events had overturned. For example the blessed isle of Chios, under Genoese rule in Thevet's time, had since then fallen under the Ottoman yoke.[63]

But the most frequent and most remarkable case, in that one sees reproduced in it the phenomenon observed above in relation to Solinus, is that of the marginal annotations that nourish Thevet's Levantine account from its own sources – notably, the *Adagia* of Erasmus[64] and the inevitable 'Ancient Lessons' of Coelius Rhodiginus. The latter, who was well known to Gregor Horst, is thus mentioned from time to time beside the very passages that Thevet had originally stolen from him. But there was no intention of denouncing him on the part of the commentator. In juxtaposing with the text of the *Cosmographia Orientis* the 'Lessons' it had originally sprung from, Gregor Horst did not seek to tell tales, but to enrich with his own contributions a book that had seduced him with all the temptations it held out for an erudite reader. Even when he had identified the literal source for a passage (as in the case of that on Asplenon, the herb curing illnesses of the spleen and causing that organ to atrophy among the animals of Crete),[65] he contents himself with noting the coincidence, without drawing from it any argument of a philological nature: 'Coelius Rhodiginus says the same in Book 4 of the 'Ancient Lessons', chapter 18, that around Cortina in Crete one finds that the livestock have no spleen; which finally the natural scientists have found to be on account of a certain herb.'[66]

As with the classical question of elephants' teeth,[67] this exact co-incidence is barely exploited beyond the slight surprise it initially occasions. Later, too, he intervenes on the subject of the earthquakes to which Crete was said to be very subject,[68] of the proverbial honesty of the islanders of Chios,[69] or of the legendary antipathy of the lion and the cock;[70] but in these examples Gregor Horst simply makes the connection with the 'Ancient Lessons', without making any issue of his annotations other than to register in an intermittent fashion, in the margins of the work, the presence of an admiring reader and accomplice.[71] The 'fiction of erudition' was thereby confirmed and reinforced by the complementary authority that the commentator brought to it.

Solinus had similarly been enriched, by his humanist commentators, with fragments from Pliny that his *Polyhistor* originally proceeded from. Now, it was Thevet's turn to be given new fame, thanks to the zeal of a Hessian doctor and to marginalia drawn from the *Lectiones antiquae* – his own primary source. In this paradoxical circulation of humanist knowledge, the latter was divided up into bundles that were then burdened with glosses enriched with the same compost; glosses that reconstituted, on slender discursive pretexts, those buried and superannuated earlier encyclopaedias. It was a process that seems to have been ruled by pure tautology. From the dawn of the Renaissance to that of the classical age, and from north Italy or the Low Countries to Germany by way of France, it was the same body of knowledge that was propagated and variously ordered according to circumstances and the public it was addressed to. But the miracle is that these circular paths of compilation renewed, in each period, the conditions of a tacit contract linking an author to his readers. The pact of erudition, reigning supreme during an age of triumphant humanism, soon degenerated into one of fiction; the pleasure of variety tended to supplant the rigour of philological inquiry. When Gregor Horst acclimatized Thevet's Levantine history to Hesse, he was not unaware that the public it was aimed at expected the refined pleasures of a discussion peppered with incidents, rather than the certitudes emanating from a magistral lesson.

3
Mythologics:
The Invention of Brazil

When [the old navigators] travelled through these unexplored regions, they
were less concerned with discovering a new world than with verifying the
past of the old. They confirmed the existence of Adam and Ulysses.
                                Claude Lévi-Strauss, *Tristes Tropiques*, trans.
                                John and Doreen Weightman

A Brazilian tropism

Bearing in mind the ten weeks Thevet actually spent at Guanabara,
Brazil occupies a disproportionate part of his work. Not only are the
Singularitez (*New Found Worlde*) of 1557 in large part devoted to it,
but the dreamed-of Antarctic France (which officially became a lost
cause on 16 March 1560) haunts, in continual regurgitations, the four
volumes of his *Cosmographie universelle* (1575). A whole book (XXI)
is taken up there by his description of the Tupinamba Indians of the
region around Rio de Janeiro. Even the *Vrais Pourtraits* (1584) select
for treatment, among the chiefs of the cannibal tribes of the Brazilian
seaboard, two 'illustrious men': namely, the redoubtable 'Quoniambec',
who captured the Hessian soldier Hans Staden, and Nacol-Absou,
'King of the Promontory of Cannibals'.[1] In the manuscripts Thevet
left at his death, the nature and peoples of Antarctic France progressively gain in importance: the late *Histoire de deux voyages aux Indes
Australes et Occidentales* represents the almost finished state of a new,
amplified version of the voyage to Brazil, the richest from a documentary point of view.[2] As for the unpublished *Grand Insulaire*, it
presented in eight maps (of which four survive today) an unpre-
cedented topographical ensemble on the regions temporarily occupied
by the French between Macaé and Angra dos Reis.[3]
   From 1557 to 1592, in effect, southern Brazil constituted for Thevet
a constant point of reference. It was an obligatory term of reference
for the description of natural and moral prodigies of the other three
continents: a means, for example, of accounting for the singular ethol-
ogy of the chameleon of Africa and Asia by comparing it with the

no less marvellous case of the Brazilian *bradype*, with its bird-like appetite.[4] Scandalizing Belleforest and the learned, who were indignant that one could take such an interest in few acres of wild country, Thevet erected his 'America' – that is, the least part of Brazil – into a paradigm of distant space.

Say, for example, it was a matter of describing the hippopotamus or marine horse that inhabits the Manicongo River in Africa. The parallel was imposed with the river whose mouth is situated 'towards the promontory of the Cannibals' and which 'those eaters of men call Toluilq, an Ethiopian word that means nothing less than Great Teeth'. Following the nomadic word, the animal was transported across the Atlantic where there were discovered, against all expectation, 'these sea-monsters, which are very little regarded by the Barbarians on account of the little pleasure and contentment they derive from them; they call them in their lingo *Naxahaquy*, that is, Good for Nothing.'[5]

Or perhaps Thevet wanted to put together, apropos of the 'Arabic gulf' or the Sea of Oman, a catalogue of 'gulfs unknown to the ancients'. In this case, examples drawn from the New World crowded on to his page to end up, predictably, at the bay of Guanabara or 'Janere', 'where I long stayed, and at the entrance to which we made our fort, for fear of being surprised by the Barbarians of the country, or others.'[6]

By virtue of the cosmographical model, which brought together extremes and marked off, according to the circles of the heavens, lateral equivalences from Orient to Occident or vertical oppositions from north to south, Brazil became the universal standard of comparison allowing one to describe the variegated unity of nature and of the nations that filled the earth. Longitudinal equivalences according to the parallels, of diagonal equivalences according to the ecliptics, were called forth by climatic symbolism, but were also translated into the structure of the continents. Already in the *New Found Worlde*, Thevet observed a similarity between the courses of the Ganges and Amazon Rivers. Just as the former 'maketh the separation of one of the Indies from another towards the East', the river named after the mythical warrior-women living without men could 'make the separation of India, America ['L'Inde Amérique'], and of Perou.'[7] From one hemisphere to the other, a natural frontier was reflected and doubled, making of the New World a mirror-image of the Old.

Between these two halves, furthermore, the common name of 'Indies' translated a solidarity of essences. In combination with a spatial metaphor he owed to the fortunate accident of Columbus – a synecdoche that took a part of the Americas for the whole – Thevet was able to constitute a particular region of Brazil into a paradigm that could be generalized out to the 'new horizons' as a whole. And at the level of customs, too, the Tupinamba would serve as a model for all 'barbarians': their material culture of the most basic order, and their superstitious beliefs, represented a convenient standard by which to evaluate the degree of savagery or civility of exotic peoples. For if it was obvious that 'these poor folk' were 'the apes of the inhabitants of the Indies',[8] the reciprocal case required that 'Indians' in the wider sense, from Persia to the Moluccas and from Arabia to Cathay, be judged by the measure of the Brazilians. From the East Indians, the comparison extended to the Arabs and the Turks, the traditional enemies of Christendom; and even to the most distant nations of Europe – from the Muscovites to the Scots, by way of the Scandinavians described by Olaus Magnus.

The legendary hostility between Turks and Arabs, which would give rise to the saying 'from Turk to Moor', could in Thevet's time go without saying, so topical was the motif becoming.[9] However, our cosmographer found it necessary to have recourse to the example of the Brazilians, who were cannibals out of a sheer appetite for vengeance, in order to give to that commonplace motif its superlative value: 'they do not love one another any more than do the Margageatz, Toupinanquins, and Tabajarres of Antarctica, whom it is impossible to reconcile to one another, such is their violent nature.'[10] It was part of the cosmographer's profession to associate in this way peoples separated by oceans, and to discover among them points of similarity or emulation.

The paradise of the Turks, which Thevet declares elsewhere to be 'very jolly' and in which the dead, sitting at set tables surrounded by affable servant-girls, freely indulged their carnal lusts, evokes the memory of a 'paradise of like order' that 'our Margageaz' imagined 'for the repose of their *Cheripicouares*, that is to say, the souls of their dead fathers and mothers'. These fortunate souls walk about 'in fine gardens, full of *Avaty*, which is millet, and of good fruits, and much *Cahoin*, which is their sweet drink'; and they 'continually gamble with their Pagès, in other words their prophets.'[11]

We discern here one of the brief sallies into comparative mythology that Thevet was fond of, and that led him to divine confusedly in the Indian myths of South America variants of 'the fabulouse picture of Melusin of Lusignan' or of Merlin the Sorcerer, or unexpected continuations of the 'conquest of the holy Grail'.[12] The merits of such a comparatism were twofold: it allowed the cosmographer to display his universal knowledge and thereby to administer a proof of his global mastery of man's diversity, since he gathered it into his hands and held it up to his gaze reaching to the extremities of the world. But beyond that, the comparison of Turks with Brazilians undeniably had another, more polemical aspect. Being invidious to Islam, the parallel had the advantage, in the eyes of Thevet and his readers, of dragging that detested religion down to level of the 'credulous beliefs of the savages' of the austral world.[13] Islamic monotheism was conflated with Indian shamanism.

The comparison held as well, imbued with a scarcely less marked disqualifying value, for the superstitions of the peoples of northern Europe. The 'idolatry of the Lithuanians before they became Christians', notably, was expressed in their adoration of 'a Hammer of monstrous height and thickness', said to have been used by the signs of the zodiac to free the Sun from a tower in which it was trapped.[14] To this striking myth, Thevet did not hesitate to compare the 'fine story' of his Antarctic savages, who imagined that three stars from heaven had once, through the will of Mair Monan, been converted into three high mountains found in their land.

The 'history of ignorance of this people' argued, moreover, in favour of their conversion to Catholicism. Thus, Thevet learned from Olaus Magnus of the beliefs of the people of Thule, who heard in the cracking of pack-ice the groans of the souls in Purgatory; and drew a parallel between this robust superstition (which had the merit, in his eyes, of breaking down the spread of Lutheranism in the Scandinavian lands), and the one that he had observed among 'those poor savages, more than barbarous, who live between the two tropics'.[15] The latter had been told by their ancestors that the plaintive song of a bird 'as big as a wood pigeon' expressed the grieving voices of the souls or *Cherepyco ares* 'of their fathers, mothers, brothers and friends who were enduring certain punishments unknown to them'. The deathly song of this flying creature – no doubt the 'Trophony' that is in question in the chapter on the Patagonian giants[16] – made it possible to attribute to those worn-out souls a prescience of the truth professed by the Catholic Church (and obstinately resisted by the instruments of the pretended 'Reformers').

These savages would soon know 'that God gives them such punishments for their sins',[17] and would then understand the reason for their posthumous misfortunes. By extending to South America the lesson of Magnus, the archbishop of Uppsala – that militant of the Counter-Reformation who, exiled to Rome, watched in terror as demons and Protestants overran his distant homeland[18] – Thevet inscribed himself within the strictest orthodoxy by demonstrating, in peoples who seemed the most distant from the divine light, 'flickerings' of the eucharistic sun that would soon blaze upon them. The Christianity lost in northern Europe was ready to be reconstituted in the southern latitudes, at the other end of the earth.

This empire of shadows that was pagan Brazil also surfaces else-where, in connection with the miracle of Versins.[19] The public tri-umph of the Holy Sacraments over a Picardian mystic evokes, in the cosmographer's memory, analogous miracles obtained through 'Agnan Hippochi', the devil of the Antarctic world; for the fact was that the savages 'were frequently delivered from him by having the Gospel read over them: such is the power of the name of Jesus over these dark powers.'

The ignorance of the Brazilians, which showed the shadows in which they were enveloped even as it augured well for their future conversion, was also translated on to the plane of material civilization. The high science of cosmography, for example, was unknown to them, and their navigation was reduced to canoe trips within sight of the coast. This model serves to describe the rough nautical knowledge of the sailors of the Caspian Sea, a people who 'observe nothing but their rutter'.[20] Deprived of compasses and nautical charts, and hoisting to their mastheads the tails of foxes to find the direction of the wind, the 'Caspians' yielded nothing in crudeness to the inhabitants of what Thevet calls, not without boasting, '*my* Antarctic France'.[21] Rough, but because of that frugal, the barbarians of Brazil observed rigorous dietetic rules; they were scandalized by the salted pork that their French colonist friends consumed without caution, shortening their lives thereby. In this, the Brazilians were made to share the dietary precepts – tinted, it is true, with Muslim superstition – of the Arabs of the Levant.[22]

Thus we see how, alternately men of nature or, on the contrary, far removed from it, the 'poor folk' of Thevet's Antarctica defined the contradictory paradigm of a non-Christian humanity. It was the very opposite of what could already be discerned in Léry's *History of a Voyage*, or in Montaigne's essay 'On Cannibals': a sort of allegorization of the savage, making him incarnate, for example, the realm of nature, primitive equality, or the leisurely freedom of an Ovidian golden age. Thevet's 'America' did not conform to any such concept. It was nothing but the sum of particular and circumstantial traits; that is to say, it condensed into itself a catalogue of 'singularities' that were irreducible and contradictory. Cruel and debauched or virtuous and hospitable, a man of honour or a 'great thief', the labels stuck on him in turn or simultaneously seem regulated by a con-stantly mobile code modelled, from detail to detail, on the particularity being thrown into relief on each occasion.

His food, his beliefs, his arts of war or medicine, define so many ways of approaching the savage and of judging him according to the contradictory appreciations he evoked. Sometimes, he comes close to the positive model sketched by Montaigne: ignorant of medicine or

physics, the Brazilians yet practise their precepts 'better than Aristotle, Averroes or Avicenna'; for 'Nature taught and teaches them every
day what is good and what is harmful.'[23] At other times, he displays
incredible stupidity; such that 'the first time they saw ships on the
seas around and approaching their coasts . . . they considered them –
never having seen such large and heavy vessels, nor their fathers before
them – as islands that were floating on the sea.'[24]

It was in all the various circumstances that made up the Indian of
Brazil that Thevet was interested; not only that universal singular one
that would later constitute the 'man of nature'. Clearly, the mosaic
and polymorphous savage of the Franciscan cosmographer was situated at the antipodes of the 'noble savage' of the *philosophes,* a pale
abstraction fleshed out by no concrete ethnographic content.[25]

We know that the modern sciences of man would originally
proceed from the latter refusal of 'singularities', which came to be
thought of as external to reason as well as to nature. The programme
of the Société des Observateurs de l'Homme, as set out in 1796 by
De Gérando, most explicitly condemned the compulsive, disordered
'curiosity-hunting' of early travellers, who were unduly attached to
the accessory and exotic and incapable of attaining to generality. It
was then that there was imposed, in place of the old rhapsodies filled
with the bric-à-brac of cabinets of curiosities, a theoretical model so
much more intolerant and reductive in its pretension to universalism:
that of a fundamentally ethnocentric anthropology.[26]

The patent lack of a unifying Reason constitutes the whole value of
Thevet's enterprise – even, one might say, in its most tangible short-
comings. The jumble of his inquiry is given as such: no finding is
withheld from the attention of the benevolent reader, not even the
most unmeditated or anodyne comment. The most violent condem-
nations – of those 'brutal' and 'bestial' savages – to be found on every
page are nearly always neutralized a line or two later by eulogies that
are symmetrical and, it seems, equally sincere or casual. Thevet is not,
of course, exempt from the political, religious or even racial prejudices
of his contemporaries (for racism, whatever has often been said of
it, was not a phenomenon foreign to the century of the great discoveries); but, since he hardly performs any selection among the
corpus of information he possessed, one would be hard put to find in
him those censures or chilly silences that emerge, for example, on the
subject of religion, from Montaigne's essay or from the *Histoire* of Jean
de Léry.

Paradoxically, it was this explosion of the Indian into a diversity of
phenomena that could make him into a universal term of comparison. The 'America' of southern Brazil is present in Thevet's world

everywhere: from Lithuania to Arabia Deserta, from Egypt to Iceland, from the Caspian Sea to Picardy. Scattered over four continents in the shower of shards from a broken mirror, the Brazilian savage was able to reflect, and thereby illuminate, the most disconcerting and scattered realities. Far from reducing the inexhaustible plurality of the universe to a demeaning unity, this patchwork of contrasts conjugated the geometrical spirit inherent in cosmographical method with the variety, at once admirable and deceptive, of collections of *rariora*. Ubiquitous yet diverse, ever present but ungraspable, he was one of the keys to cosmography as it was practised by Thevet.

In this sense, the cosmographic revolution brought about by the man from Angoulême – in the sense one intends when speaking of a Copernican revolution (although, of course, cosmography kept to the geocentric schema of Ptolemy) – consisted of privileging a marginal place (Brazil) in relation to a traditional centre (Europe, or its age-old opposite, the Mediterranean Near East). Since his descriptive system had a double nature – being ordered within the geometrical canvas of a map, yet always returning, in the final analysis, to the sacrosanct principle of autopsy – he had to choose between these (roughly antipodal) poles: the Levant of his pilgrimage in 1549–52, and the southerly far west of his voyage to Guanabara in the winter of 1556. The Orient of ruins and stereotypes that Thevet inherited from a centuries-old tradition, in which his only scope for innovation lay in the rather boring mode of denial, would soon be supplanted by the new splendour of an unheard-of Brazil, whose ample richness is affirmed over and over again in his successive works. Displacing a worn-out Levant – whose redundant description was ever more cut off from its living sources in his peregrination, and drawn from easily identifiable compilations from that of Paolo Giovio to that of Sansovino[27] – the Brazilian tropism gradually came to govern the spaces of the Thevetian *mappa mundi*.

Being physically dispersed in this way over the terrestrial orb, Brazil also had a similar extension through the language of its aboriginals. One finds in Thevet – at least prior to the unfinished *Grand Insulaire* – neither exotic dictionaries nor bilingual *colloquia* (as in Cartier or Léry);[28] but erratic phrases placed in new situations, or strange sounds uttered in inappropriate contexts. The word *margajat*, for example, which designated an urchin, would pass into the current usage of the classical age, no doubt by way of Thevet or one of his imitators.[29] But the cosmographer also uses the nickname *Gentil Morbicha* for Luther who, he says, had 'forbidden all his disciples and adherents, on pain of being rejected from his Church, to enter into disputes with Catholic ministers'.[30] The term *morbicha* referred, in Tupinamba society, to the wise old men who would gather in the 'long house' to deliberate about a war to be undertaken. Again, in a passage of the *Cosmographie universelle* Thevet does not hesitate to compare them to the 'lords consulting in the Senate at Venice; such is the gravity they display, and the modesty of their consultations.'[31] Elsewhere, he describes a *morbischia-ouassoub* or 'great king' strutting, 'stark naked as he was' and with a club on his shoulder, in front of Villegagnon and his followers, who had barely landed from their ship.[32] It is patently comic to imagine Luther as a cannibal chief and nudist, presiding from the heights of his hammock over the debates of his confederates. But it is by no means certain that a reader not forewarned would have appreciated the irony of the situation – this obscure and barbarous term being explained only sixteen pages further on.

Even more unexpected, and almost incomprehensible, is his conclusion, in the collection of *Hommes illustres*, to the chapter devoted to Robert Gaguin, the author of a *Compendium de Francorum gestis* ('Collection on the deeds of the Franks').[33] Joining to a eulogy of this representative of early humanism that of Rodolphus Agricola, of 'Dr Jason Maynus of Milan, the reformer of the law',[34] and of the 'satirical' Jean Lemaire of Bellegem, Thevet introduces, apropos of the latter, an onomastic 'allusion' that seems most incongruous. This author of the *Légende des Vénitiens*, so 'marvellously piquant' in the 'sallies' of his sharp pen, appeared to our prosopographer a peerless historian in his master work, *Illustrations de Gaule et singularitez de Troye*; 'For although some have wanted to babble, if one bothers to consider attentively the discourses of this Mair, very few rarities will be found of the things that have happened in Christianity up to his time, that he does not touch upon.'[35]

This pun or *annominatio* between *Lemaire* and *ce Mair* would border on the absurd for anyone who did not know that *Mair* (or *Maire*) was the generic name of the civilizing heroes in Tupinamba cosmogony.[36] Thevet himself gives the name the primary sense of a 'transformer' – that is to say, a magician; and he devotes no less than three chapters of the *Cosmographie universelle* to narrating in detail the mythical avatars of Maire-Monan, Maire-Pochy and Maire-Ata.[37]

When one knows that Jean Lemaire tried, by maintaining the primordial indistinction between myth and history, to elevate the latter through the former, one cannot fail to be struck by the relative appropriateness of the allusion Thevet proposes. Revealing, by the intermediary of the pseudo-Berosus of Annius of Viterbo, the fabulous origins of the two sister monarchies, German and French, since the Deluge and the sack of Troy, Jean Lemaire was in a sense comparable to the caraïbes (*karai*), or Brazilian prophets. They united in their person the qualities of both shaman and demi-god, and appeared to the Indians as reincarnations of the primitive *Maires* – especially of the chief among them, Maire-Monan, the author of infinite metamorphoses of himself and of the most diverse objects.[38] Their 'theology' (of which Thevet notes that it was 'couched, not in writing, but simply in the memory of everyone'),[39] also narrated, apart from the said 'transformations', myths of origin – not that of a princely dynasty (and for good reason), but of inventions such as the cultivation of maize and root crops, or the use of fire.

Again, Jean Lemaire's historical *bricolage* is not without recalling the refinement of the courts of Burgundy and France and, in addition, the 'savage mind' of the Tupinamba as it was minutely transcribed by André Thevet, and coloured with ironical asides. Using whatever means came to hand, accommodating the most disparate traditions within the same body of doctrine, the poet of Marguerite of Austria and Anne of Brittany brought together, in his project for a political and cultural concord for Christian Europe, the materials for a history of its common origins. Such was the enterprise that Thevet, embroidering the lesson of the *Mairs*, enlarged on in his *Singularitez* or *New Found Worlde* of 1557.

Polydore Vergil and the savage mind

His translators served André Thevet well. In two cases at least, they brought to light his most roundabout intentions and discerned his least apparent sources. Thus the German Gregor Horst outlined the relation of kinship or even dependency that existed between certain chapters of the *Cosmographie de Levant* and the 'Ancient Lessons, of Coelius Rhodiginus.[40] Similarly, the Englishman Thomas Hacket found that Thevet's mode of investigation in trying to explain the indigenous societies of Brazil and their primitive culture was close to the method followed by Polydore Vergil, the historian of England and honorary citizen of that realm, in his treatise 'On the Invention of Things'.[41]

In his 'epistle dedicatorie' to the *The New Found Worlde, or Antarctike*, addressed in 1568 to Sir Henry Sidney, Hacket develops at length the list of inventors of arts and sciences without whom man would still be naked, barbarous and brutal, even a slave. He passes in review the progressive institution of the calendar, from the ancient Egyptians to Julius Caesar by way of Numa and Romulus; the imposition of civil laws by Isis; the teaching of agriculture by Ceres; and the beginnings of navigation under the influence of Minos, Neptune or his father Saturn, or again of King Erichthas – all pagan monarchs, to whom should of course be preferred Noah, as a biblical patriarch.[42]

This summary of Polydore Vergil, whose name is mentioned in connection with the invention of the calendar and who systematically followed a Euhemerist mode of explication, ends with a eulogy of navigation and an express call to plant colonies overseas. Deploring the pleasures of Capua, where his contemporaries lazed, 'almoste abhorring to heare the name of travell or payne', Hacket exhorts them to follow the example of Alexander, a fervent reader of Homer and of the exploits of Achilles (if we are to believe Plutarch). In making available to the English public the book of the honourable traveller 'Andrewe Thevit',[43] Hacket offered it at the same time the means of putting that noble enterprise into practice.

In a few pages, the translator and editor has thus outlined the literary and political design of his model, and been able to adapt it to contemporary English realities. In fact (and despite their discretion on this point), the *Singularitez* were originally involved with the colonizing activity of Villegagnon, for which they constituted an especially seductive sort of prospectus. The programme of conquest sketched out between the lines of the work was now adapted to English designs on the New World. Furthermore, in summarizing Polydore Vergil at the threshold of a description proceeding from a similar philosophy of origins, Hacket clearly discerned the ideological point of the *Singularitez*. No more than Gregor Horst, indeed, was he playing the role of philologist or historian of literature. What he perceived, from Polydore Vergil to Thevet, was not a linear and unequivocal filiation but a community of preoccupations; an analogous method, that resorted to a comparison of cultural traditions between various peoples in seeking to arrive at a general model and to base on the latter the superiority of Christian Europe.

The comparativism that the *Singularitez* set in place was the fruit of the obscure labour of Mathurin Héret, a bachelor of medicine and classicist, known for his translations of Alexander of Aphrodisias and Dares Phrygius.[44] When Thevet's book was published in December 1557, indeed, Héret brought an action for recognition of authorship. He obtained a judgement in which all royalties would be paid to him, although Thevet would remain nominally the author. It was to Héret, no doubt, that had fallen the task of stuffing the account of Brazil with references to Greek and Latin authors. Whence the parallel construction of some chapters of the book: after a precise ethnographic exposition come examples drawn from antiquity, by the convenient mediation of Pliny (who is sometimes quoted) or of Polydore Vergil (who is never quoted). It might, for example, be a matter of evoking the good memory of the Indians, who never forgot the names of their French guests once they had heard them the first time. A list of champions of the mnemotechnical art – 'Cyrus King of the Persians, Cyneus the legate of King Pyrrhus, Mithridates, Caesar' – is immediately produced, taken from Book 2 of Polydore's treatise 'On the Inventors of Things',[45] to form the conclusion of the chapter. Another example: the quasi-monastic tonsure of the Indians, which was justified by concerns of a military order (hair would offer the adversary an easy grip during hand-to-hand combat). This calls forth, as an explanation and justification, the precedent of Theseus offering his hair to the gods at Delphi; and the somewhat more adequate one of Alexander, who had the heads of soldiers of the Macedonian phalanx shaved in order to make them completely ungraspable.[46] The source, here again, is the opuscule of Polydore Vergil, which saved Thevet or his secretary the bother of going to Plutarch's *Lives*.[47]

In a similar fashion, the chapter devoted to the *Pagès* and *Charaïbes*, who were respectively the healers and prophets of the Tupinamba, concludes with an inquiry into the invention of magic that again attaches the Brazilian corpus to a recurrent thematics of the origins of the arts and sciences.[48] Or again, the same might be said of the 'visions, dreams and illusions, that these miserable "Ameriques" have', who are entirely under the sway of Satan's empire. The nocturnal persecutions inflicted on them by the 'wicked spirite' Agnan, and the premonitory dreams visited on them on the morning of a battle, can be compared to the two kinds of divination discerned by Polydore Vergil.[49] In turn, a series of themes are illustrated by the example of the Brazilians: those of primitive warfare between naked and Herculean warriors armed with clubs,[50] of the laws of marriage,[51] the birth of commerce,[52] the invention of the saw,[53] the first habitations and the origin of architecture,[54] or the mode of subsistence of early man.[55]

The reference to Polydore Vergil is ubiquitous in this Brazilian sequence, and underwrites the transition from the post-humanist *bricolage* of the *Cosmographie de Levant* to the unspeakable, historically shallow novelties of the *Singularitez de la France Antarctique*. Almost all the chapters that form the core of the latter yield a binary sequence associating some 'ancient lesson' with a motif of an ethnographic order. The gesture is repetitive, incessantly effecting the reduction of the unknown to the known. From a primary foreignness, observed and related, we are brought back to the familiarity of a text many times read and interpreted. Its end result is to substitute for Brazil and its cannibals Homeric Troy, the Scythia of Herodotus, or the Golden Age of Ovid and Virgil. The enterprise of Thevet–Héret merely systematizes, in this way, a tendency observable from the earliest accounts about the New World.

Since the *Decades* of Peter Martyr d'Anghiera, the discovery of the 'new Indies' had completed the renaissance of literature by working, paradoxically, towards the same end. The distant in time and the distant in space both defined a common territory in which classical culture could once again find itself at home.

Or, more precisely, the parallel it offered with the customs of antiquity and its Euhemerist traditions conferred on the society of the cannibals the dignity of an object of science. Erudite references, given for the most part at second hand, were doubtless intended to place European man and the naked, anthropophagous American on the same footing. But above all, they authorized the latter to penetrate into the field of 'histories'. By a double process of heroification and moralization, the 'most cruel men in the universe' were admitted into rivalry with the illustrious men of Plutarch: Theseus, Lycurgus, Solon, Caesar.[56] Thevet's enterprise of integrating American 'glories' into the traditional corpus of the great men would culminate in the *Vrais Pourtraits*: Quoniambec, Paraousti Satouriona, Paracoussi the king of La Plata and Nacol-Absou, minor princes of America, joining there, in the same format, the figures of Alexander, Caesar or François I of France.[57]

This grafting of antiquity on to the New World by the efforts of Héret also engendered a secondary relationship, tending in the opposite direction. If America was justified by reference to the ancients, Brazil in return explained to Europe its own origins. Thus, the American ethnographic treatise could appear at the same time to be a manual of European archaeology. Our own earliest forefathers went naked, fought each other with tooth and nail, and perhaps went so far as to eat their defeated adversaries; they, too, were ignorant of the arts of the forge or the rules of marriage; and their houses were of woven lath, unless they preferred the shelter of caves. All these propositions are found in Polydore Vergil. Passing in review the West's civilizing myths, the humanist from Urbino had proposed for each 'invention' (language, religion, the calendar, the art of war, agriculture, architecture, navigation, commerce, pottery (*puterie*), etc.) the name of one or more heroes, demigods or prophets. This composite mythology was destined to encounter the founding myths of Tupinamba religion, as reported by Thevet. The kinship was evident: in both cases, the history of humanity was reduced to that of the great initiators. From then on, a transfer became possible from one continent to the other: Noah or Daedalus could take the place of Maire-Monan and Maire-Pochy, for example, as usurpatory (and necessarily fictive) figures.

This brutally reductive reading in fact continued one which Polydore Vergil had earlier defined, when he denounced the lies and impostures of ancient paganism in order to exalt, by contrast, the privilege accorded exclusively to Christian Revelation. For example, Pliny's catalogue of inventors in Book VII of his *Natural History* had served to map the geographical distribution of merits among the various peoples of the Mediterranean basin, who each in their way could claim to have contributed something to man's well-being. To the Phoenicians was due the honour of inventing navigation by the stars; to the Egyptians, that of inventing the alphabet and weaving; to the Phrygians that of the four-wheeled chariot. The Cretan Daedalus had discovered the arts of 'building', the Phoenician Cadmus the extraction and smelting of gold, the Theban Tiresias the art of divination from the flight and cries of birds.[58]

For Polydore Vergil, on the contrary – following in this the *Contra Apion* of Flavius Josephus, one of the main sources for his compilation[59] – the ultimate truth of the Bible and the Chosen People consigned to oblivion the competing pretensions of the other nations. Cain had invented cultivation with the plough long before the birth of Ceres, a goddess who was thus brought down to the rank of a humble mortal. Noah supplants Bacchus, and Moses does not wait for Hercules to invent the art of war. The voice of the patriarchs and judges thenceforward drowns out the (on the whole harmonious) concert of different national claims. Far from effecting a synthesis between Christianity and paganism, the treatise 'On Inventors' subordinated the complementary diversity of the pagan traditions to the unity of a totalitarian truth. Had not Polydore Vergil, after all, recognized in Christianity the finest of all 'inventions'; the initiative for which, as for the rest of creation, belonged to God alone?

We know that to his first three books, published from 1499, Vergil twenty-two years later added another five devoted to the origins and institutions of the Church. As Denys Hay has observed,[60] this did not modify his initial project. The humanist from Urbino had from the outset pursued an apologetic enterprise based on mythographic comparison. In this, he was following a trail blazed by Josephus, and favouring Jewish culture; one then taken up on behalf of Christianity by Eusebius, in his 'Evangelical Preparation'.

Imitating it in turn, Thevet showed himself faithful to the spirit of the treatise 'On Inventors'. Thus, when he needs to recall the ages of primitive humanity, he refutes in the name of Genesis the fables of the poets Virgil ('in the first *Georgic*') and Ovid, who supposed that 'all men universally on the earth have lived like as do the brute beastes'.[61] Eden is not the Golden Age, and the work of a gardening Adam already separates man from the indigence in which the animals wallow.[62] Then, apropos of 'Abell, and of his first fruites that he offered [as a farmer] to God' – a confusion with his brother Cain that was intended, no doubt, to glorify the noble vocation (profession) of agriculture,[63] Thevet completes the destruction of the fictions of paganism by insisting that man worked the earth from the beginning. Thus, the savages of Brazil, who cultivate manioc and maize, are (in spite of their ignorance and uncouthness) part of a labouring humanity, and share in its promise of salvation. The lesson of Polydore Vergil here militates in favour of monogeneticism, at the same time as establishing a hierarchy of knowledges and cultures.

The common front formed at this juncture between Holy Scripture and Indian reality broke down, however, when it came to the question of religious beliefs. The exclusive privilege retained by the Judaeo-Christian tradition in the face of pagan mythology was confronted with the 'histories' (*comptes*) of the Amerindians concerning their origins. Thevet's 'Singularities of the New Antarctic France' were pleased to underline the contrast: the 'blind brutality' of the poor 'Ameriques' burst through, for example, in the credence they accorded to their shamans and their vain 'sorceries'.[64] This countervailing eloquence was exposed to the Christian reader's view in order to encourage him to follow the straight and narrow path of Grace. In the same way, in his chapter on the visions and persecutions of the malign spirit, the author concludes by inviting the reader to turn to the Bible; he thus confounds, in the same reproval, the idolatry of the 'ancient Gentiles' and the respect the savages had for their '*Pages* or *Charaibes*, which is to say, halfe Gods'.[65]

From these American myths of origin, then, there was little to be drawn other than what accorded with Revelation – namely, the belief in a universal deluge of water and fire, which was widely attested throughout the Brazilian and Amazonian region.[66] As for myths of inventors, at most there would be credited to the Indians those that had something to do with exotic crops. It was enough to refer to their 'traditive' account of the invention of manioc cultivation – that poor substitute for bread, of which they were deprived; for this they were essentially indebted to Maire-Monan, the civilizing hero of the Tupi.[67] For the rest, it was necessary to return to the teachings of Scripture.

This archaeology of Europe by way of an interposed America was thus intended less to support the idea of a continuous progression from one to the other, than that of a fundamental rupture between two ages that everything had conspired to separate: that before and that after the Christian Revelation. The Tupi had not, until this time, been admitted to the era of Grace. Their separation from 'truth' was expressed, in a very concrete manner, by their manifest poverty and barbarity. Nudity and cannibalism were its most tangible signs, but also their ignorance of elementary techniques such as cultivation with the plough, the smelting of iron and forging of metals, the equestrian arts, and firearms.

In the end, then, the savage served as a foil for European Christians: as an example of a humanity endowed, notwithstanding its scant respect for the commandments, with divine election and with the certainty of a future redemption. He offered a radical counter-example for medi- tations on living one's life according to God (as, later, for Jean de Léry).[68] Useful for the very separation he represented, the Indian – who would gradually be condensed into the figure of a creature living according to the flesh – initially had emblematic value within the moral and theological discourse that enveloped him. But with Thevet, thwarted by the heroic vision of history inherited from Plutarch, and giving way moreover to a fascination with the fragmentation of the diverse, that discourse failed to achieve coherent expression; it remained in the state of a plausible, yet non-exclusive, beginning.

In it, the *bricolage* effected by Héret remains visible: it shows through in the schematism of the two structures summarily fitted together and running in parallel from chapter to chapter. It is Polydore Vergil, no doubt, who has the last word: he is always mobilized at the last minute, to close any question. His genealogical apparatus supplants the meagre mythology of the Tupinamba, but assimilates it incom- pletely. The fine 'histories' of the Tupinamba were progressively less well mastered, to the extent that Thevet's Brazilian *oeuvre* was am- plified during the successive stages of the *Cosmographie universelle* and *Histoire de deux voyages*, so that before long they would explode the precarious construction assembled by Thevet's scribe Héret in 1557 for the *New Found Worlde*. From 1575, with his polyphonic suite of myths narrating the creation of the world or the metamor- phoses and death of the culture-hero Maire-Monan, the universal deluge and the different ages of humanity, Thevet composed a veri- table *Mythologics*. Not content, for example, with reporting the story of the deluge and the two brothers, he juxtaposes two distinct vari- ants of it.[69]

No doubt we should see in this a desire to use up the last vestiges of his material; and in this sense, the cosmographer reveals himself a consummate *bricoleur*. But one also discerns in this iterative schema an obscure prescience of the principle that a myth is defined by the 'set of its variants'.[70] The book 'On Inventors' progressively gives way to the oral redundancies of an indigenous cosmogony into which creeps, from time to time, the outline of a comparison. Thus the transformations of Maire-Monan recall for Thevet the enchantments of Circe,[71] and the inheritors of Maire-Ata, who had to undergo a series of trials to prove the origin of their blood, are associated with

the one in question, was adapting an Indian reality to Polydore Vergil, or Polydore Vergil to the imperiously mythopoetic logic of the Amerindians. Later, after the *Singularitez* or *New Found Worlde* (a work contemporary with the colonial enterprise, and in which a missionary perspective can be discerned), the didactic or even apologetic character of Thevet's writing would be mitigated, and the savage thought of the Tupinamba would be less bridled by sententious chapter-endings that unfailingly brought their cultural alterity back to the fold of Christian universalism. Then the mythic *bricolage* of the Brazilians would encounter that of the cosmographer, itself 'entangled in imagery',[82] and from those concrete unities or 'singularities' would be constituted a science located half-way between percept and concept.

4
Mythologics II:
Amazons and Monarchs

We finde by the histories, that there are iii. sorts of *Amazones*, differing only in places and dwellings. The most ancient sort were in *Affrica*, among the which were *Gorgonists*, that had Meduse for their Queene. The other were in *Scythia*, neere to the river of *Tanais*, which since have raigned in a part of *Asia*, neere to the river of *Hermodoon* [Thermodon]: and the third [quatrième] sort of *Amazones*, are those which we do treat of.

André Thevet, *The New Found Worlde*
(*Singularitez de la France Antarctique*)

A mêlée of naked warriors

In the *Singularitez*, Thevet's recourse to Polydore Vergil allowed him to establish a basis of common origins between his tableau of American customs and the realities familiar to Europe. In what amounted to a reductive strategy, it involved subordinating the Indian alterity to a historical schema that privileged – by anteriority and by an inflated authority – the Christian tradition. This conscious enterprise would engender, it is true, perverse effects: instead of purely and simply integrating the culture of others, it would end up as a sort of hybridization.

To this ambiguous, incomplete phenomenon of contamination I would give the name 'pseudomorphosis' – the term Erwin Panofsky applies to the subterranean transformations undergone by the iconographic types transmitted from antiquity to the Renaissance, after the long but fecund sleep of the Middle Ages.[1] Here Love, as if emerging from a dream, is blindfolded, and Old Man Time awakes armed with the scythe that had been associated with him in apocalyptical imagery. Thevet's American Hercules, with his head and loins covered by feathers, his club tapering like an oar, and brandishing the trophy of a freshly cut human head and members, arose from a similar order of surreptitious, insidiously fantastic images.

There were, however, certain differences from iconological pseudomorphosis: the *bricolage* here was instantaneous; and it proceeded from a calculated operation in which the element of play was, as we have just seen, not inconsiderable. The iconic lapsus, the elastic *Witz* (joke), here arose not from an involuntary and collective confusion produced by fortuitous transmissions over the course of centuries. It resulted, on the contrary, from a deliberate *bricolage* and a punctual initiative. Ultimately, the synthetic image resulting from this abridged process did not acquire the same degree of universality as the figures of Love, Death or Time studied by Panofsky. Thevet's Brazilian Hercules, as represented in his *Vrais Pourtraits et Vies des hommes illustres* of 1584[2] – or the feathered Gaul that the poet du Bartas introduces in his *Seconde Semaine*, banqueting under ash trees and waiting for acorns to drop from oaks[3] – were private creations whose fortune barely outlasted that of their authors. When these American pseudomorphoses did attain some perennity, it was because their semantic usage became specialized. Thus, an Indian Artemis, riding an armadillo and wearing a diadem of feathers, could designate only America in collections of emblems in the Renaissance and classical age.[4]

With these reservations we can proceed to study, through some particular illustrations taken from the corpus of the great discoveries, the way a mutual contamination of mythologies on either side of the Atlantic came about. The observer's gaze (in)formed the reality he described and in which he found, in an eminently fortuitous way, the barely transposed image of his own inherited obsessions. Reciprocally, the Indian stories he transcribed and inflected according to his own criteria parasitized, in their new variants, European representations, creating, by a process of grafting and montage, new symbolic objects: an equatorial amazon and an Adamite, cannibal king.

We might begin with a plastic example of this phenomenon of crossover contamination. In the *Singularitez* (and then in the *Cosmographie universelle*, for which the engraving was reworked in a larger format),[5] Thevet illustrates a scene from a savage war, in which contorted athletes grip each other in hand-to-hand combat. Wooden clubs and showers of arrows are not enough to express the aggression of these vengeance-seeking barbarians: we see them sinking their teeth into each other's calves and arms. One of them has seized the perforated lip of his adversary in order to pull him close and strike him more conveniently. The engraving, in these two versions, illustrates the combat of the Margageats and the Tabajares, enemy tribes of the same Tupi population in Brazil: 'they grab and bite one another with their teeth in any place they can get hold of, including their perforated lips.'[6]

The 'hideousness, mingled with entertainment' emanating from this ferocious spectacle[7] is translated by the exaggerated lines that link the image, in mannerist style, to the parts of Thevet's work where the anatomy of the Indians is exposed in its most Dionysian aspect.[8]

The scene was, in fact, inspired by a pre-constructed model, the inevitable collection 'On Inventors' by Polydore Vergil. The latter tells us that, before the use of arms, primitive soldiers fought 'with fists, claws, and by biting with their teeth'.[9] By combining the club of Hercules, which represents a more advanced stage in the arts of war, with that primitive fighting in which man used only his natural arms, Thevet's engraving brings about a synthesis in which observation of the Indians plays the least part. Only in a few revealing details (such as the club shaped like a flattened spindle, the ring of feathers worn on the haunches, or the pectoral crescent) can the tableau be associated with the actual military behaviour of the Tupinamba. The bodies with elongated muscles, the oval shields, the bows and lances would fit as well in chapter 10 of the second book 'On Inventors', on the origins of military art, as in this manual of ethnography before its time, the *Singularitez*.

Yet the composite nature of the image was precisely what commended it to the attention of Thevet's contemporaries, and in the first place to the artists who were inspired by it. Imitating, it seems, the initial version of 1557, the 'Melee of naked Warriers' inserted by the engraver Étienne Delaune into his series of twelve *Combats et Triomphes* exploited the topical value of the original image.[10] In the woodcut from the *Singularitez* he had found the form of a frieze or bas-relief which he enlarged and enriched with secondary figures. To accentuate the sculptural aspect of the full-face bodies, he substituted for the landscape in the background a black base, in common with the rest of his images. Placed after a triumph of Bacchus and a theriomachy with unicorn and griffin, elephant and dragon, wolf, bears and lions, and before a combat of Lapiths and Centaurs, Thevet's mêlée of Tupinamba warriors was thus gently integrated into the sequence of glories and violence that made up pagan mythology.

The only disparate element among these muscular warriors, who otherwise resemble the Lapiths at the wedding of Hippodamia, was introduced by attributes that allow us to recognize authentic Tupinamba. The aquiline nose, the leer of a mouth deformed by the wearing of a labret (lip-ornament) and offering the adversary an ideal grip, could not, any more than their hirsute chief, distinguish them from the cortège of combatants found in other scenes. On the other hand their arms – those tapering clubs that Delaune could have sketched from examples in cabinets of curiosities in Paris[11] – or their dress, their diadems and ostrich feather boas,[12] constitute irrefutable proof of their belonging to the Brazilian region. Furthermore the artist threw over the shoulders of one warrior a feather cloak very similar to that which one sees in Thevet, in an engraving showing the funerary 'banquets and dances' of the Indians.[13] The motif of a combatant lying on the ground and biting his adversary on the calf, thigh or arm is repeated after Thevet three times; notably on the right side of the composition, where the supine figure with a bear-like face reappears in reverse. This detail recalls the folk tradition of a 'savage man' awakening, dishevelled and hairy, at the end of winter,[14] and joins that, in two parts of the scene, showing a hare-lip (*balièvre*, or *bec-de-lièvre*) holed and distended by the victor's finger, in evoking somewhat insistently the primordial stage of war before the invention of arms, described by Polydore Vergil.[15]

Closer to these fighters reduced to their natural defences of tooth and nail than to Thevet's Tupinamba warriors, the savages of Étienne Delaune are still depicted using stones and uprooted trees. They represent, thus, the lowest stage of development in the arts of war. Comparable to the drunken centaurs in the following plate, they do not know how to transform into human tools the brute objects that nature provides them. Or rather, they use their rudimentary artefacts, arrows and clubs along with unworked materials such as stones and trunks. When, at the end of the duel, they no longer have any weapon in their hands and are reduced to the mouth and fingernails with which they 'bite and scratch each other',[16] the gap separating them from animality seems indeed small.

Now Delaune, playing the game of serial contamination, amused himself by transplanting into neighbouring plates iconographic elements drawn from the Brazilian image. We have seen the symbolic link tying the latter to theriomachy, in which animals confronted with each other according to their mutual antipathies (lion with griffin, elephant with dragon, dromedary with horse), also had to contend with hunters armed with spears and clubs. Later, on the contrary, the artist is pleased to emphasize the discordance between this manner of fighting and refinements of armour and military costume in the ancient style. The tenth engraving of the series of 'Combats and Triumphs' shows a mêlée of cavalry and infantry, whose oriental scimitars, round shields decorated with a flaming sun, tufted helmets or corsets and leggings are combined with an uprooted tree brandished in the manner of a ram, and with biting. In this compositional fantasy, where the gestures and attitudes of savage warfare are deliberately mixed up with the armaments of Romans and Ottomans, the remnants of barbarous hand-to-hand combat result in a studied effect of disparity.

With the recurrence of quasi-animal violence in an image where we see, joined to that of the janissary, the panoply of the perfect Roman legionary, the whole series acquires a sort of fusional coherence: the stages of the progress of military art are confounded, its codes of representation and its references scrambled at will. The mythological register, with its triumph of Bacchus, its combats of Lapiths and Centaurs, is degraded into the carnivalesque, in this scene where men, women and children ride asses and fight each other with ladles, spindles, sickles and flails.[17]

From chivalrous heroism to a farmyard parody of it, and from human warfare to animal ferocity in its monstrous avatars (griffins, dragons and unicorns) or its more familiar ones – cock-fights, dogfights or the squabbles of ducks, among the rural brawls evoked above – Delaune's polemological series declines variations on a theme; though not without distinguishing between the successive stages of a sequence. In this fine disorder, unified only by its graphic medium, frontiers are abolished between animal and man; between brute nature with its strange productions, and the refinements of civilized war – between the primitive violence of a world at its origins and modern rituals of fury. The abundance of hybrid, legendary or mythological creatures, from griffins to centaurs, betrays that fundamental indistinction and illustrates, in a tumult of bodies and forms, the Dionysian excesses of military activity. The cortège of battles is introduced by a triple evocation of the ancient triumphs of Bellona (plate 1), Victory (plate 3) and Bacchus (plate 5) in such a way that the mannerist suite can absorb into itself, as one of its components, a Herculean and feathered cannibal. The latter was thus 'mythologized' and thereby admitted into a synthetic imaginary in which the reference to antiquity appears predominant but not exclusive.

Delaune realized in images the effect Thevet had tried to achieve by recourse to a mythography of 'inventors': a composite alloy, in which the new gained a dignity equal to the ancient by being integrated into a formal, traditional framework. But his difference is none the less evident. Where Thevet had tried to organize and classify by basing himself on distinctions drawn by Polydore Vergil, Delaune mixes and confuses, and produces a fascinating ambiguity. The *Singularitez* had situated American barbarity on a double scale, chronological and hierarchical, in an attempt to measure the whole distance separating Christian Europe from the Brazilian savage who was destitute of all practical or spiritual knowledge. The graphic creation of Delaune, on the contrary, results in an osmosis; one that is emphasized both by the persistence of a style and by the repetition, from one plate to another, of a composition in scrolls and flowing arabesques.

Thevet's engraving had fixed, in the battle scene between Margageats and Tabajares, a moment in human history. It was an illustration –

condensed, no doubt, but coherent – of the primitivism invented a century earlier by Piero di Cosimo in two cycles of paintings devoted, respectively, to Vulcan and Bacchus.[18] Delaune, while drawing on this same fund of icons of the human origins, seeks to show neither a dramatic progression nor a historical rationality. Certainly he does not ignore the existence of correspondences between, for example, the myth of the centaurs and the brutality of the first ages; and on occasion he even extends them by juxtaposing the hand-to-hand combat of cannibals with the brawl at the wedding of Hippodamia. But these are simply elements of a learned play, in which the spectator is invited to discern in the pleasant variety of things offered to his view the confused stages of a process, and a set of diffuse cultural allusions. The whole is resolved, for the pleasures of the eye, into a frieze with a regular rhythm, by turns alarming and joyous. What is gained on the aesthetic level corresponds to what is lost in meaning.

A more systematic pseudometamorphosis was achieved in the *Divers Pourtraicts et Figures faictes sus les meurs des habitans du Nouveau Monde* ('Various portraits and figures of the manners of the inhabitants of the New World'), by the Poitiers engraver Antoine Jacquard.[19] Posterior, no doubt, to the expedition of Razilly to the Maranhâo in 1614, this series (a frontispiece and twelve oblong plates, each containing four subjects enclosed by arcades) is related by its pronounced mannerist style to the 'Triumphs' of Delaune. It is of smaller scale, and its fifty or so figures of feathered cannibals, rosy-cheeked children, buxom women and muscular, long-limbed men tend to evoke the art of the goldsmith rather than that of the sculptor.

Having had access to the cabinet of curiosities of Paul Contant (a 'Poitiers apothecary' and friend of the bourgeois Jean Le Roy, to whom the *Divers Pourtraicts* were dedicated),[20] Antoine Jacquard obtained the exotic attributes with which he furnished his gallery of figurines: fruits such as the pineapple (2,2; 4,4; 5,2); maize and calabashes (5,3); or fauna bordering on the teratological, such as the toucan (13,4), a serpent with forked tongue (13,3), or a flying fish (2,1) with the tapering beak of a bird (13,2). Rings, diadems and crowns of feathers, religious and military items such as maracas, bows and clubs, containers made of tapir skin (9,1; 11,3 and 4); not forgetting the peaceful presence of a hammock (8,3) – these all allow one to identify in his images the 'inhabitants of the New World'. To these revealing details are added quarters of human flesh which, distributed over plates 6, 9, 11 and 12, designate the cannibals of Brazil or the Antilles.

Since the time of Thevet or Delaune, whom Jacquard used among other documentary sources, models of exotic representation had been multiplied and diversified by iconographic collections of the magnitude of the *Grands Voyages* of Théodore de Bry[21] and, above all, by the vogue for cabinets of curiosities in which natural prodigies and native artefacts were displayed pell-mell. Antoine Jacquard was thus able to illustrate a variety of objects unknown to his predecessors. To take here only one example, the club, we note that besides the Herculean example – a simple tree-root, barely squared and still showing its nodules (6,3) – three distinct types coexist in this later American suite. The *iwera pemme* or Tupinamba sword-club, which Delaune had borrowed from Thevet and which proliferates (on the authority of Staden and Léry) in the third part of de Bry's *America*,[22] still predominates. But we also find with Jacquard a Gê axe shaped like an anchor (6,2),[23] and the large *boutou* of Guyana, with its flat edge enlarged in the manner of a painter's spatula (9,2). This same 'sword' from Guyana is found in a water-colour in the anonymous *Histoire naturelle des Indes* in the Pierpont Morgan Library, which dates from the beginning of the 1590s. Entitled 'Hindes de Ihona' (from the Spanish pronunciation of Guyana), it represents an Indian attacking another, who is knocked off his feet and whose smashed head streams with blood.[24]

Most interesting of all is the way Antoine Jacquard brought together, by associating them with a traditional symbolics, these fragments of a disparate ethnography, whose objects are taken from all over the Amerindian region. Does not the hairstyle of the child brandishing a flying fish (2,1) recall the 'Huron' hair of the Canadians? The little children represented in the frontispiece as exotic cherubs, the savage women bearing offerings, arise from the same plastic register as the naked warriors who execute, shield and club in hand, a sort of Pyrrhic dance (8 and 10). The demonstration of various attacking and defensive gestures, the catalogue of bellicose postures (10 and 11), still belong to a gestural topics inherited from antiquity. Jacquard's disjointed frieze, its figures isolated in the intervals of porticoes, obeys a disposition as 'classic' as in Delaune's series of bas-reliefs.

But Jacquard went further, in seeking to draw the naked, mannerist body of the American in two directions. On the one hand (in a deliberate allusion to the *Fabrica* of Vesalius), it tends towards the anatomy text: flayed bodies and skeletons are alternately shown in two of the friezes (6 and 7) wearing bandoliers of feather boas and with bows and maracas in their hands. On the other hand, it tends towards allegory: a skeleton puts to its mouth a horn with a flag showing a gorgon's head, as it swaps the Indian club lying on the ground at its feet for a scythe; the latter transforming the Brazilian's 'anatomy' into an emblem of Time and Death (7,1).

By turns, the engraver dressed up his cannibal with exotic attributes and stripped him down to a skeleton: one of these figures, a flayed body, wears its skin like a scarf, while a woman next to him, in the same walking pose, wears nothing more than a tuft of feathers over her bones (6, 3 and 4). In playing on the Indian body in this way, comparing its nakedness to that of bones, Jacquard made it support multiple and contradictory meanings. A new figure, hitherto foreign to emblematics, received contents as disparate as death and Dionysian barbarity.

It is no accident that Jacquard's engraved suite coincides in places with the apparatus of anatomical texts of the time. As André Chastel notes, the medical discourse of Vesalius and his successors was framed by a moral discourse on human fragility and the inevitability of death, which was read spontaneously as 'a gigantic *memento mori*'.[25] A comparable alloy of wonder and terror was cast in the spectator's mind by the 'inhabitants of the New World' – a gallery of savages that opened and closed with the gesture of an offering of fruit and venison and, in between, deployed in colours of red and gold the horrors of cannibalistic butchery accompanied by grotesque dances and capers. It was a meditation at the same time joyous and macabre.

The reduction of the savage to a familiar model, and his integration into a frieze in the ancient style, did not exorcize the fascination exerted by a foreignness in which the horrible mingled with the bizarre, the admirable with the repugnant. The 'Combats and Triumphs' of Delaune and, perhaps more so, the *Divers Pourtraicts* of Jacquard bear the traces of an effort, even an obsession. An effort to comprehend and restrict that alterity, at once dangerous and convenient, into which were thrust with impunity the secret dreams of the artist; and the obsession of a desire that was pushed back to the human origins and banished to the body of the barbarian, but whose image surreptitiously returned to haunt the spectator.

The Amazons of Brazil would also be integrated into the categories of ancient knowledge and arise, as such, from an age-old science, to reproduce the same contradictory movement of differentiation and identification, of horrified exile and disquieting proximity. By a new pseudomorphosis, developing simultaneously on the levels of aesthetics, ethics and epistemology, the myth of the female archers was to be durably incarnated in the Indian women of the equatorial forest.

The four sorts of Amazon

In his *Singularitez* or *New Found Worlde* Thevet, as a traveller returned from Brazil, proudly proclaimed that the three sorts of Amazon described by the ancients were now to be augmented by a fourth.

From the earliest times, Europeans had known of the Amazons of Africa (among whom were counted the gorgons), of Scythia (of whom Herodotus wrote in his book on Melpomene: IV.110–17), and of Asia (those living on the banks of the River Thermodon). Now, the 'American Amazons' – a 'fourth sort of Amazon'[26] – had come to complete the lacunary descriptions of antiquity. The grid was filled: each continent, around Christian Europe (but, of course, excepting the latter) had its Amazons. On this point, as on many others, traditional knowledge found its fulfilment in the discovery of a world whose existence it had little suspected.

The conformity of these new Amazons with the old ones is easy to demonstrate, if only by the exact tautology that encased the reasoning writers used. It was, in fact, by means of a quite conventional descriptive schema that these women, as ferocious as they were 'marvellable' (*esmerveillables*), were represented; so it is not surprising to find that such an imaginary construction corresponded in all points with the premises of the description. Thevet, moreover, reveals in spite of himself the circular logic presiding over the erection of the myth, in a clumsy preamble: 'Some may say, that they are not *Amazonists*, but as for me I judge them suche, seeing that they live even so, as we finde the *Amazonists* of *Asia* to have lived.'[27]

The behaviour of Amazons was based, as we know, on a schema of inversion. They devoted themselves to all the activities normally reserved for men, beginning with hunting and war, and despised on the contrary those that were generally the domain of their own sex, such as housework or horticulture. All that remained to them was the education of their children – of their daughters, that is, since boys were killed at birth. The sexual division of labour was thus preserved, but a systematic inversion was effected, from the masculine pole to the feminine. This inversion was associated with perceptions of a violence carried to paroxysmic levels, as expressed in the castration or murder of male infants; it was as if the inversion of real society could engender only a nameless barbarity. Such was the topos of a 'world upside down', so frequently illustrated in the sixteenth century, and a particular variant of which would be the 'Amazon complex'. It stigmatized the present disorder by imagining a hyperbolic reversal and appealed – by way of this recourse to the scandalous figure of inversion – for a return to the traditional order of things.

From this point of view, the myth of Amazons did not escape the rule: one consequence of the conquest of the Americas would be to make them vanish, as a monstrous and fantastic anomaly; and this thought is implicit in the majority of writings on the subject. Thus Thevet laments the fate of the 'poore people' who fell into their hands and found 'no great consolation among these rude and savage women'.[28] The cruel tortures he attributes to them, illustrated by a most eloquent engraving, consisted of hanging their (male, of course) victims up by their feet and shooting arrows at their naked bodies 'above ten thousande times', while a fire was lit under the unfortunate men's heads.[29]

We know how ubiquitous the ferocious sisters of Penthesilea were said to be at the time of the great discoveries. Basing himself on a study of several dozen Renaissance travel accounts, Georg Friederici found traces of them in at least eight distinct areas of the New World: in Brazil, first of all (where their legendary memory is permanently inscribed on the land, designating the world's largest river basin); as well as on the plateaux of Chaco and the south-west of modern Brazil, in Guiana, in the Antilles (where Christopher Columbus picked up rumours of them on his inaugural voyage of 1492), in Colombia (where Fernan Pérez de Quesada described the gynocratic state of a woman cacique, Jarativa), in New Grenada, in Nicaragua and, finally, in Mexico, where an Isle of Women was found off the coast of Yucatán, discovered by Pedro de Grijalva, the immediate predecessor of Hernán Cortès.[30]

It was this too perfect conformity with the ancient myth that aroused the suspicions of the chronicler Peter Martyr d'Anghiera. Evoking, in the fourth decade of his *De orbe novo*, Grijalva's reconnaissance of the Yucatán coasts, he displays some disbelief in the account (which, however, he reproduces) of those women 'of evil customs' who, in their earliest youth, cut off one nipple 'so as to be able to draw their bow with greater agility'.[31] 'I think this is a fable', he concludes in the manner of a denial; though having already mentioned the furtive love affairs of the warrior women with men who visited their islands for as long as it took to couple with them.

It seems too simplistic an explanation, however, to attribute the origin of this universal fable only to the fantasies of European men. In fact, it seems that the Western myth of Amazons encountered, in the virgin lands of America, a very similar Indian myth, that the legend imported by the conquerors found itself confirmed, in a way, by the natives' own beliefs. Indeed, if one looks closer, one notices that the legend of Amazons was always attributed to a native source. It was on the basis of reports given by the Arawaks of the Antilles that Amazons followed cannibals[32] into modern travel literature, in the first account of Columbus in 1492.

The modern mythologics of Claude Lévi-Strauss, moreover, attest to the presence of such a myth across several Apinayé, Carib and Warrau versions, which link it (contradictorily) to ideas of seduction by honey and to the origin of tobacco.[33] Thus, the confinement of 'tobacco-crazed women' to an island had deprived men of the drug, rendering them incapable of elevating themselves by smoke towards the supernatural beings, a disorder which had its inverted complement in the culpable gluttony of a 'honey-crazed girl', who ceded too manifestly to the seductive power of nature and was immediately punished for it.

The myth of Amazons – articulated here with the risk of a regression of culture towards nature, a mortal risk expressed in the antithetical terms of a dearth of tobacco and an excess of honey – was already perceived in such a way by certain Renaissance travellers. It suffices, for example, to read carefully the two chapters André Thevet devotes to amazons in his *Singularitez* or *New Found Worlde* of 1557. Alongside the traditional repertoire of traits borrowed from antiquity by way of the 'Ancient Lessons' of Coelius Rhodiginus, from the *Cornucopiae* of Niccolò Perotti, or from one or another of the Latin 'dictionaries' of Calepino and Estienne,[34] one notes a coherent set of characteristics whose common effect is to define a regressive stage of humanity. The brutality of those naked and castrating archers is conveyed not only by the barbarous torture described above, but in their very way of life: fishing, hunting and the gathering of 'rootes and some good fruits that this land bringeth forth'.[35] These are their only resources – excluding, it seems, gardening and agriculture, at least if the 'rootes' mentioned are consumed, unlike manioc, raw. Their habitat is equally rough: their 'litle lodgings and caves against the rockes' evoke an architecture much more rudimentary than that of the *malocas* or 'long houses' that housed up to a hundred individuals among the Tupi-Guarani tribes described elsewhere by Thevet.

It is true that the androcidal women's habit of burning their victims entirely would appear to argue against such a rejection of Amazons to the realm of uncultivated nature. But this excess of cooking paradoxically joins up with the technological insufficiency just remarked on. The Amazons were no more able to cook than they were able to live in society with members of the opposite sex, except to communicate with them in furtive embraces 'at some time secrete in the night, or at some appoynted time'.[36] It was a shortcoming fundamentally linked to the very absence of what, elsewhere, represented the most 'vituperable' of Amerindian customs: namely, the practice of cannibalism. Their inability to love, serve and eat their masculine partners illustrates, in a particularly flagrant way, the retreat of these Amazons into a world before all culture.

The symbolism of the myth is expressed also in terms of topography: the city of women most often appears in association with an insular location. Whether it is a question of the Warrau myth concerning the origin of tobacco,[37] or of Thevet's description of the Amazons' villages,[38] the island bespeaks reclusion and the separation of the sexes. Only from time to time, for example, when their annual mating is held, does the feminine island open itself up to masculine assailants. It is in this way that the Amazons perpetuate their kind, in spite of their visceral hatred for the opposite sex. A woodcut in the *Singularitez* shows a somewhat less idyllic episode of this intermittent bodily encounter of the two sexes. Besieged on their island by *canoas* full of men, who attempt to disembark while shooting arrows and brandishing clubs, the naked warrior women with raucous voices and flying hair stand their ground against the aggressors. The shells of marine tortoises of which their ramparts and shields are made repeat *en abîme*, as it were, the insular figure that forms a sort of emblem of these women without men.

'Woman is an island': the association, as eloquent today as in the Renaissance (and recently taken up in an advertisement for perfume), is all the more pertinent in that it underlines the irreducible isolation of the other sex. Situated in warm seas where the very wind is fecund (in the words of Pigafetta's old pilot-book), or brought before Orellana and the conquistadors who came from the Andes by way of the immense arms of the eponymous river, the islands of Amazons were by vocation unreachable. All that could come of a skirmish between Spaniards and Amazons, such as Thevet relates in the mock-heroic mode,[39] was a fiery exchange of volleys of arrows and cannon shots. 'Enchanted isles', 'Isles of the Blest', were so many fantasmatic objects that were kept from the navigator (like the feminine body) during wanderings that extended over whole years. It was in the quest for those ungraspable drifting islands that was consummated the feminization of the archipelago.

At this point in our analysis, it is no longer clear what should be apportioned, in matters mythical, to the native or to the European observer. Like those of the fountain of youth, of Eldorado or of the Land without Evil, the myth of the Amazons of America seems to be exemplary among cases of 'imaginary and mythical contamination', of which François Delpech has proposed a formal history, and which cannot be reduced to a simple play of borrowings and influences.[40]

The Amazons encountered by Francisco de Orellana in 1540–1 while travelling down that river from the Andes represent rather well an antithesis to the Indian women of the Atlantic seaboard, such as would be so precisely described by the French in the following decades. Thevet noticed this relationship when he stated, on the subject of the military practices of the Tupinikin of the Rio de Janeiro region, that 'the women folow their husbands to the warres; but not for to fight as doe the *Amazoness*, but for to minister to their husbands foode and other necessaties, requisite in the warres.'[41]

In reality, in other words, those who could perhaps have played the role of the legendary Amazons were far from making war; they pursued the most humble and arduous material tasks. The unequal sharing of duties between the two sexes, during long military expeditions sometimes of six months or more, merely extended the division of labour observed among the same peoples. 'The women travail more than do the men', notes Thevet elsewhere, 'that is to wit, to gather rootes, make meale, drinkes, gather together the fruites, dresse gardens, and other things that appertaineth to householde.'[42] The men, meanwhile, quite nonchalantly devote themselves to the pleasures of hunting and fishing, or make the arrows and bows that will be their only baggage during their periodical moves. An engraving in the *Singularitez*, linking in the same picture three distinct scenes (a tribal march, the use of the petun (tobacco) cigar and fire-making with a drill), reveals the contrast between the women, bending beneath the weight of heavy baskets of portage goods and surrounded by children, and the lone man walking ahead of the group with a lighted cigar in his mouth, his knife in his belt and a bow in his hand.[43]

Four centuries later, Claude Lévi-Strauss would formulate comparable observations on the Nambikwara in the Brazilian interior; here, however, it was the men who concerned themselves with gardening and, inversely, there was a class of leisured and warrior women who lived in intimacy with the chief. It remained true, however, that during the seven months of the year when manioc was scarce, the Nambikwara's subsistence rested almost entirely on the women: their gathering of fruits and roots was actually a more reliable resource than masculine hunting.[44]

In the *Cosmographie universelle*, Thevet brutally expresses such domination of one sex by the other: 'Their horses, mules and baggage carts are the women, who are responsible for providing munitions and carry them on their shoulders with the men; without regard for any aspect whatever of the infirmity of that sex.'[45]

No doubt the hypothesis according to which the American Amazons were inverted Indian women (who mirrored, perhaps, a convergent negativity towards the Christian women on the other side of the Atlantic) no more constitutes a generalizable model than does the one that would reduce the myth to the misinterpretation of a distant reality. But at least it has the advantage of including the latter in a certain way, since the discourse on Amazons, whether emanating from native or European sources, always expressed the feeling of a fascinating and scandalous difference. If one talked about Amazons both here and there, it was because one saw, through the mixture of wonder and horror that they aroused simultaneously in the narrator and his audience, an obsession touching on the very foundations of the society to which one belonged. What would become of the latter if, one day, women broke away from the society of men, or even exterminated them? As with Herodotus (whose Scythians, however – who normally incarnated barbarity – end up being confused with the Greeks, and play exactly the same role in the face of absolute alterity as do these Amazons),[46] Thevet's gaze united the Tupinamba and the Christian Frenchman in a common opposition to the insular hunting women of the great river.

On the other hand, it is worth noting that the relationship of his 'Amazonia' to existing, ethnographically recorded societies is not ordered by a mechanical inversion. There exist from the outset, in societies under masculine domination, disquieting signs that expose them to a switching of sexual poles, and even to a secession of one group from the other. The indignation Thevet showed on the subject of the division of labour among the Tupinamba should be understood in this sense: the rude treatment to which the native women were subjected (with, moreover, their full consent) seemed so revolting to him only in that it scoffed at the difference between the sexes. By assigning to such weak creatures the arduous tasks that should by rights be those of the stronger sex (in accordance with a natural law easily confused with a divine ordinance), the Indians authorized that floating of limits and fluidity of distinctions that brought with it, on the rebound as it were, the full amazonian subversion.

The notion of independent and fierce Amazons, of course, contradicted the attitude of the submissive and peaceable Indian women that, as Thevet kindly points out, could be compared to beasts of burden. But it is no less evident that they merely drew out and questioned that attitude, since the Tupinamba women already transgressed, by their way of life, the normal bounds of their sex. In the end, and if one is correctly interpreting the presuppositions in Thevet's vision of the Other, the revolting alterity of the 'rude and savage' Amazon women was present in a latent state in Tupinamba society, which offered from the outset an all-too-perfect negation of it. Between one society in which women gravitate, without regard for the supposed 'infirmity' of their sex, to the most back-breaking occupations, and another in which they dominate and supplant men in all things (or almost), there is no rupture, but only a veiled continuity.

A passage from the *Cosmographie universelle* dealing with Imaugle or Imangla, the island of women situated in the Indian Ocean somewhere in the region of Ceylon,[47] again evinces such a menace. An allusion is slipped in, in the course of a refutation of the myth of Amazons, to 'one who made the booklet entitled The Most-Marvellous Victories of the Women of the [New] World'. This was Guillaume Postel, the author of that opuscule (Paris: Jean Ruelle, 1553) and moreover, as we know, a friend of Thevet. He had not only exposed the 'admirable excellences and facts of the feminine sex', but pretended furthermore to demonstrate 'how it was necessary for it to dominate the entire world'.[48] This apocalyptic triumph of women would be effected by the mediation of the 'most sacred Mother Johanna', latterly Postel's companion in Venice where she ran the Hospital of Santi Giovanni e Paolo for the sick and indigent, and who had revealed herself to be the new Eve, the mother of the world and the bride of Christ. Announcing, with the brilliant miracles accomplished during her lifetime, the reign of the Spirit in which all things would be restored to their original state before the Fall, she would bring to its end the vast cyclical movement of Redemption begun fifteen centuries earlier with the incarnation and death of the crucified Christ. As a privileged witness to this revelation of pre-eminent importance for man's future, Postel had noticed the perfect coincidence between the advent of Mother Johanna and the great discoveries that allowed the Christian Word to be proclaimed to the entire universe. It was in particular the 'inferior hemisphere' that would benefit from this late unveiling – that is to say, not so much the austral hemisphere as that half of the world to the west, beyond the oceanic regions where the sun went down.[49]

One can understand, in such a context, the role of Amazons, whose sudden ubiquity in the century of the great discoveries had essentially, for its theatre, that 'temporal, feminine, occidental, descendent, infernal, mutable part' of a world at last rendered in its totality to human knowledge.[50] The revelation of 'very great kingdoms in southern Africa and in America around Peru', now ruled by women,[51] had eminent value as a sign, and left no doubt as to the sovereign mission devolving on to the feminine sex in the divine plan of restitution. Amazons, past and present, naturally took up their place at the threshold of Postel's book of 'Most Marvellous Victories', and it was this liminary mention that Thevet, in his *Cosmographie*, judged as inopportune as it was misplaced.

For with Postel, the question of Amazons had become a metaphysical one, and Thevet foresaw the dangers of such an idealization. Here we leave behind the realms of scientific heresy for those of heresy pure and simple. Hence the vigour with which the cosmographer reacted against these dangerous feminists, even if it meant denying his previous assertions.

However, one could not so easily dispose of half the world, be it ever so inferior and mutable (to use Postel's terminology). Thevet found this out in spite of himself, and the three chapters of his *Cosmographie universelle* that seek to make an end of the tenacious ubiquity of Amazons, who were spread over Asia Minor,[52] the Indian Ocean and Brazil, demonstrate on the contrary the myth's perennity. The Island of Women situated in the waters around Ceylon, the fabulous Imaugle, has nothing to do with an island of Amazons, warns Thevet at the outset; yet by insensible degrees the treatise reverts to the commonplaces of the Amazonian topic. These were, in the first place, the 'accumulated troops of women' along the river banks travelled by solitary navigators, who appeared identical to the female warriors of legend;[53] and then the dexterity of these women in 'handling the bow, and drawing adroitly with it'.[54] It was true that the islanders had husbands and children, but the husbands lived apart on another island, and had only intermittent relations with their legitimate spouses. This function of distributing geographically the tasks devolving on to one or other sex seemed to have originated with the binary archipelago of Imaugle and Inébile. The women were employed in gardening and in building huts made of woven palms and twisted branches. The men, who ordinarily lived from fishing, devoted themselves as well to hunting and, 'seeing themselves without the burden of women or children, to many fine braveries'. But such a symbolic dichotomy very quickly becomes confused as the women of Imaugle, themselves assailed in their island following the defeat and massacre of their husbands, have to behave like authentic Amazons and make war.

The point was that the two islands corresponded, in their peaceful coexistence, to two types of economy that were too evidently antagonistic to be able to subsist in a durable equilibrium. While the island of men was consumed by continuous warfare, the island of women, for a long time spared, was ruled by laws of conservation and of a strict reproduction of forces. After fecundation on Inébile, the wasteful island exposed to all ravages from the outside, the nine months of gestation were spent among the gardens of Imaugle. The energetic separation grew apace between these two sexed poles of the archipelago, in spite of the regular transfer of male children by the end of their seventh or eighth year to the men's island. In order for the demographic equilibrium of the two islands to be maintained, it was necessary for certain girls and women, 'the strongest they could select', to be metamorphosed into men, playing the role of bellicose and vindictive males and, in extreme cases, attacking enemies in the manner of 'new Amazons'. Then the formerly closed society of the women of Imaugle was opened up to distant exchanges, in bloody maritime expeditions whose victims were the 'poor inhabitants of the island of Bazacate', whom they put to death, in reprisal, 'cruelly and poorly'.[55]

We see that, in spite of all his initial denials, Thevet has recast in the double insular paradigm of Imaugle and Inébile the schema typical of Amazonia – all except for the burning-off of the nipple, which in any case he had already, in the *Singularitez*, refused to believe. The myth of Arabian origin to which he shows his debt in this chapter of the *Cosmographie universelle* has ended up triumphing over his superficial attempt at rationalization. Or rather, its insular structure has carried the day over the reassuring image the cosmographer tried to give of those 'domestic and affable women' who returned to their island once pregnant, and faithfully kept the home for a father who was occupied elsewhere with warfare.[56] In conserving the island theme, which is here much more than a simple topographic support but the structuring form of the myth, Thevet could not help evoking the insoluble question of the division of the sexes; and of that which is inextricably linked to it, the foundations of civil society. The Amazonian obsession that turns around, at the end of his story, the author's proposition, re-emerges in the final analysis in order to impose – against all evidence – the truth, not of a history, but of a symbolic fable.

The final example borrowed from the *Cosmographie universelle* again dramatizes this 'fourth sort of Amazon' that Thevet had discussed in the *Singularitez* or *New Found Worlde*. Uncharacteristically, he does not here make previous or competing cosmographers responsible for the Amazonian mystification, but valiantly turns the blame on to himself: 'and I am indeed a fool, to have fallen into the error of believing it.'[57]

It is true that such an admission is straight away tempered by the collusion he attributes to the savages, who had 'led themselves to believe in fine reveries' and transmitted them to over-credulous travellers. But Thevet, despite the polemical rage that inspires him, is not without regrets for the beauty and strangeness of the myth he henceforward sets about exploding, and to which he had himself so eloquently contributed. Hence these two full folio pages, devoted to a people who did not exist!

This new chapter on the Brazilian Amazons also responds to a necessity of another sort: in 1575 Thevet still had at his disposal the woodcut engravings that had illustrated the *Singularitez*; notably, the two plates illustrating – how vividly! – the bravery and cruelty of the island women. In a concern for economy that had manifested itself earlier (when, in the *Cosmographie de Levant*, the author said he did not want to 'waste the effort' of some coins reproduced for a

proposed collection of medals),[58] he now reused in their existing state
two engravings that had become inadequadate to his polemical intentions. They had tended to prove, by the active virtue of the image, the
existence of the fictive warrior women; whereas now they coexisted
on the same page as a long text proving the contrary. In order to save
his plates, which he paradoxically continued to regard as documentation, Thevet was thus led to employ various subterfuges. In the first
place, the engravings were accompanied by discreet 'headliners' (marginal notes) restricting their meaning: 'The women defending themselves against their enemies', or 'The cruelty of those warrior women'.
Thus it is nominally not a question of Amazons here. The image was
not, thenceforward, in absolute contradiction with the commentary,
which suggests that (without being, properly speaking, Amazons) the
women of the Margageats, a wild and cruel people, 'fight alongside
the men when there is a need.'[59]

Thevet's dodge would become one of the favourite explanations of
historians of the age of discovery to account for the extension of the
Amazonian myth to South America.[60]

It remained to justify the separation of the sexes, too explicitly
displayed in the woodcuts that ranged men and women in two camps
confronting each other. An ingenious fiction (which curiously rejoins
the destiny of Imaugle) was devised to demonstrate that this war of
the sexes was only apparent. If the Indian women, normally laborious
and submissive to their husbands, were metamorphosed into redoubtable archers, it was as a result of a purely fortuitous coincidence
– of circumstance. While their husbands fought on another front, and
believed them to be secure back on their islet, an enemy stratagem
took the women by surprise, and obliged them to defend themselves.
The arms they then turned on the enemy were, initially, of a specifically feminine order: 'menaces, shouts, and the ugliest and most hid-
eous faces and grimaces that can be imagined'.[61]

The absence of their husbands raised their ardour in the struggle
and also explains, according to Thevet, their use of turtle-shells for
protection, for want of the natural rampart formed by the head of the
family. As in the legend of Imaugle and Inébile, it was thus through
a lack of men, and not their own will, that these women temporarily
turned themselves into Amazons.[62]

A reading so constantly reductive, re-establishing step by step the
superiority of the masculine sex (while retaining the myth's most
sensational details, such as would hold the reader's interest), becomes
even more flagrant in the commentary to the second engraving. Hitherto the naked and armed Indian women have been described as 'poor
women' who, 'in the absence of their husbands tried to hold on to

their goods, their lives and their children'. Now they are degraded into a howling, bloodthirsty horde to whom the men, their husbands, or rather the chiefs of the horde, hand over their prisoners when they return from a victorious expedition, 'as do deer-hunters, who want to get a dog accustomed to the scent and odour of venison'.[63] This remark is thrown into relief, in particular, by the fact that it concludes the whole of the foregoing episode on Amazons.

But despite the abrupt nature of this irrevocable fall, it is clear that an ambiguity remains. Just before the closing phrase, an incidental remark betrays the return in force of the Amazon myth. For a moment, Thevet lets himself be caught up again in the heroic and brutal dream of the warmongering island women: 'And I can assure you that there is no nation, from one pole to the other, that is more fierce than this feminine sect.'[64]

Thevet's terms 'nation' and 'sect', here, can logically mean only the Amazons whose existence, precisely, has just been contested. Thus, he corrects himself with a final sentence which brusquely denies the autonomy of the female 'sect' in question and, thereby, the extraordinary nature of these cruel women archers. As good spouses, they obey their husbands right down to the refined tortures they inflict on their captives. For it is 'to please their husbands', indeed, that they rush for their quarry and display a wealth of inventiveness in making the enemies of the family die slowly over a low flame. So everything returns to normal: the reign of men is not threatened. And yet a dream has taken place, one so insistent that it needed the acrobatics of a commentary to surround and subjugate it. In the end, it left behind two fascinating – and, in a word, incomprehensible – images of violence.

The case of Thevet seems symptomatic of the perennity that is, so to speak, consubstantial with the amazons. Their singularity is maintained by preterition, by means of a well-known rhetorical mechanism.[65] An apparent denial covers, in fact, the continuous transit of information as fabulous as it was 'marvellable'. Jean de Léry would not be mistaken in 1585 when he mocked, in the third edition of his *Histoire d'un voyage faict en la terre du Brésil*, Thevet's Amazons entrenched behind their huge turtle-shells, who were 'no more a novelty in these lands than in those of yore'.[66] His riposte might seem in bad faith, if it did nothing but reproduce the principle of negative authority that is so prevalent in Thevet's own work, and that Léry takes over almost without realizing it. By virtue of such a principle, the author being refuted – *Thevetus refutatus*, as the table of authors puts it in the 1586 Latin edition of Léry's *Histoire*,[67] – authorizes digressions into frankly imaginary realms. Léry, in his *Histoire*, does not give any more credence to fables of Amazons in the 'Land of Shilly-shally' than does Thevet in his *Cosmographie*.[68] Yet he could evoke in great detail this legend in his own book once the supposed stupidities of his predecessor had given him the scientific right, and a convenient pretext, to do so.[69]

The myth of the Indian monarchy

The sort of hybridization effected in the myth of Brazilian Amazons is again manifested, though to a lesser degree, in the case of the naked and anthropophagous king. Resulting from the convergence of a poorly understood ethnographic reality – the episodic role of the war-chief among Tupinamba tribes – and the transfer of an institution familiar to European colonists (the reign of a monarch with unquestioned prerogatives), this myth does not, however, show the same purity as the previous one. Whereas the legend of fierce archer-women was largely abstracted from any historiographical rationality, a large element of calculation entered into the construction of the Indian monarchy as it was elaborated by sixteenth-century Europeans.

Expressing an anxious interrogation of the foundations of society, the obstinate spectre of women without men has never ceased to describe – contradictorily – the fears that weave their way through the relations between the sexes. But that of a figure at the same time repulsive and fascinating, the myth of a cannibal king, was – even if it was originally founded on a comparable ambivalence – soon transmuted into a practical model: for on its adoption by, notably, Thevet (or its rejection, by Léry) depended the very viability of the colonial enterprise. For a discoverer or conqueror, it was above all a matter of finding the right person to talk to. Isolated individuals encountered at random on unknown shores offered no kind of guarantee unless they spoke in the name of some authority – unless, in other words, they bore the power delegated by some chief or king. More generally, power over others is possible only on condition that they belong to a set of social structures.

By contrast, this political myth tended to lose all its use-value for a simple observer of native customs, an ethnographer before his time, however disinterested a Renaissance Westerner might be when confronted with the intact image of his own origins. The model of the naked monarch fell into gratuitousness, for want of the imperious and fatal necessity that gave the Amazonian fantasy its value. In him, the savage mind was from the outset combined with the cold, calculating reason of the soldier or the merchant.

Situated midway between its obscure origins and an impoverishing elucidation, and alienated by aims that were foreign to its internal economy, the myth of an Indian monarchy none the less constituted an illuminating case of the imaginary contamination between cultures. What it lost in formal purity it gained in ideological efficacity as, in this example, mythologics were articulated with a political project. Fiction was transported from the poetic and religious domains to that of political ends. It anticipated a real situation; and it was their refusal of the coming history – one of acculturation, of decline and fall – that would motivate, among Léry and his readers, the destruction of that myth.

There was a figure who, in the course of successive versions given by Thevet in his Brazilian reportage, from the *Singularitez (New Found Worlde)* to the *Histoire de deux voyages*, steadily grew in amplitude: that of the 'demi-giant' Quoniambec, the Tamoio chief of the region of Rio de Janeiro, whose imposing stature was said to have no parallel except the extent of the territories under his jurisdiction.

The historical existence of this person seems to be well attested; he should be identified, no doubt, with Konyan Bebe, who held the Hessian archer and mercenary Hans Staden prisoner for about ten months.[70] According to the respective eye-witness accounts of the German and the Frenchman, Quoniambec combined a voracious appetite (did he not claim to have eaten five Portuguese?) with a somewhat exhibitionist habit of boasting. In both the *Wahrhaftige Historia* and the *Cosmographie universelle*,[71] he is represented in the same posture: strutting proudly before an audience of fearful subjects in the space of a *maloca*, declaiming with great strides the glories of his military exploits. The 'ambulating harangue' of the Tomoio chief assumes in the two works, however, opposing significations. For the prisoner Hans Staden, destined to serve as food for the Indians, Konyan Bebe was an intractable murderer abandoning him to the sadistic games of his men, before whom he had to jump with bound feet and endure their jibes and blows.

With Thevet, on the contrary (who was among the French allies of the valiant 'king'), the portrait, even though fearsome, was intended to arouse sympathy. And that, by two apparently contradictory means: situation comedy and epic aggrandizement. A highly colourful figure, Thevet's Quoniambec takes on an undeniably Rabelaisian aspect: having drunk some vinegar mixed with water (the only alcoholic beverage available to Villegagnon's colonists), he belches loudly and then bursts out laughing at these digestive manifestations, as eloquent as they were unexpected. Upon which, he beats his chest, 'shoulders and thighs'[72] and informs the assembled public of his horrible desires, which would immediately reduce the strongest and most able of his 'Margageat' enemies to a 'hash stew' if he had him in his grasp. Covered only by the feathers of his diadem and the rings of paternosters (prayer-beads) at his collar, the king was 'totally naked'; yet that did not inhibit him as he loudly narrated the litany of his prowess, in a voice so frightful, Thevet assures us, 'that you would scarcely have heard thunder'.[73] The 'audience' the French captain had accorded to this 'most venerable' of his Indian partners thus came close to sliding into outright comedy. However, the elementary rules of diplomacy forbade that such a face-to-face encounter be mocked; all the more so because the savages' anger might become murderous, or at least disastrous for the handful of colonists, whose survival depended on native supplies.

This oscillation, making of the scene Thevet evokes a model of diplomatic ambiguity, expresses with singular acuity the relationship of European to Indian. The latter, who is ignorant of established usages (even that of clothing oneself), is the object of an implicit contempt. Hence the derision that attaches to Quoniambec's gesticulating, jokingly demonstrative nudity. But in a parallel fashion – since the relation to the Other appeared under the sign of necessities of a strategic order – the danger was that the interlocutor might become aware of the contempt in which he was being held, and refuse peaceful exchange. Hence too, therefore, the sort of symbolic compensation whereby Thevet raises Quoniambec to the rank of a monarch with eminent qualities. Physical qualities, to begin with: 'tall and big-limbed, with a height of some eight feet', the Tamoio chief is said to possess such strength that 'he could have carried a barrel of wine in his arms.'[74] To underline the hyperbole, Thevet does not hesitate to picture him a little further on carrying on his shoulders two light cannon, which he fires simultaneously at his enemies.

In crediting the Tamoio chief with this Herculean prowess,[75] the cosmographer was acting the part of the perfect courtier towards the unlikely monarch his pen had engendered. Indeed, by virtue of a style often bordering on the encomiastic he progressively metamorphosed the Brazil of the cannibals into another Europe. Thus, the house or *maloca* that was festooned with the heads of enemies taken in war and ritually devoured was honoured with the title 'palace' from the moment the 'king' had stayed in it.[76] For such a sovereign were pronounced, when he was brusquely snatched from the affection of his subjects by an epidemic, the funerary rites reserved for the brave: 'Had he lived, he would have achieved great things with the support of our people.'[77] The physical qualities of the hero have their corollary here in a moral elevation in which the appetite for vengeance gives

way to a sovereign carelessness about death, and boastfulness is transmuted into a disinterested love of glory. To complete the fiction,
Thevet imputes to the crude warrior, in a most unlikely way, a devo-
tion worthy of the reforms of Trent: 'He took such great pleasure in
watching us as we made our prayers, that he himself sank to his knees
and raised his hands to heaven as he had seen us do.'[78]

Clearly, this series of commonplaces defining a perfect gentleman
and showing in Quoniambec an accomplished prince assumed the
function of protocol. Quoniambec is a model 'noble savage', suscep-
tible of conversion to the Christian faith and of thereby becoming a
dependable cog in the wheel of the colonial enterprise that was getting
under way at the time under the auspices of Admiral Coligny. By
inflating a eulogy concentrated on a single chiefly figure, the latter
could be made at the same time into a model and a relay: the model
of a partner whose virtues conformed to the chivalric and feudal
ideals of the European captains, and a relay between the civilized and
savage worlds. That relay was incarnated to perfection by the hybrid
link in which a riotous and risible nudity was linked to the majesty
of a powerful body and a noble soul.

Now this astonishing and complex portrait, whose ideological and
political implications I have attempted to isolate, would be taken by
Thevet's adversaries as a particularly eloquent example of the latter's
vanity and stupidity. Jean de Léry's *Histoire d'un voyage* takes for its
favourite target, in a polemic mainly directed against the king's
cosmographer, the 'feral' Quoniambec.[79] Insisting, for example, on the
obvious unlikelihood that two pieces of artillery could be mounted
on the shoulders of a naked man, and simultaneously discharged
without that living support sustaining the slightest scratch, Léry quite
rightly scoffs at the 'impostures' and 'stork-stories' falling from the
pen of the prolix 'cosmographer by royal letters'.[80]

This violent attack would have for its reply a chapter of the *Vrais
Pourtraits* in memory of the valiant Tamoio chief. Stung by his adversary's charge, Thevet reinforces his previous statements and accentuates
the hyperbolic nature of what now amounts to a eulogy in the most
traditional form. In his gallery of illustrious men Quoniambec, ranked
alongside the Caesars and Tamerlane,[81] appears as the prototype of
the new humanity discovered beyond the oceans, one surpassing 'in
many things' the ancient world.[82] The hitherto manifest barbarity of
the cannibal king is now, to say the least, mitigated; in this whole
chapter there is never any mention of either his anthropophagy or his
elementary costume.

His idealization used as its go-between a periodic style and a bevy
of unexpected metaphors. Thus, the 'frightful' Quoniambec, 'garlanded

with the most exquisite rarities appertaining to both the body and the mind', appears furthermore to 'radiate many virtues' that incline him, for example, towards things religious. The Latinisms of a vocabulary rather tainted with pedantry make of the subject's 'gigantine procerity' the tangible incarnation of the 'eminence of degree' in which he was placed. Here Thevet conforms to a traditional mode of thought that perceived in the physical a reflection of the moral. A king worthy of the name could not be of common or inferior stature. Montaigne would attribute similar reasoning to a cannibal captain encountered in Rouen, who was amazed that the child-king Louis IX could reign, despite his fragile appearance, over a squadron of solid Swiss guards.[83] Further, it seems that Thevet took up, here, an old hypothesis concerning the origin of monarchy, which the cleric Jean de Meung had earlier expounded in the *Roman de la Rose*. The latter would be cited, again in the context of a comparative ethnographic description of the peoples of America, by the lawyer Marc Lescarbot at the beginning of the seventeenth century:

> They chose a sturdy peasant, big of bone
> The largest limbed and tallest there was,
> To be of all the seignor and the prince.[84]

At the origin of feudal society and of the monarchical principle that crowned it had been a common need for protection. It led to the election of the strongest, soon aided in his defensive task by an escort of sergeants. Effectively, according to Thevet, Quoniambec was the 'stoutest' of the Tamoio, the one whose 'eight feet' of height assured him of incontestable superiority over the members of the group. Gigantism appeared in this sense to be the sign of supreme dignity, at the same time as being the indispensable attribute of the naked monarch. Without this 'gigantine procerity', how would one distinguish him among thousands of his subjects, equally naked and worthy?

It is true that Thevet invests with royal signification the least of the rare ornaments adorning the chief's powerful anatomy – as becomes particularly evident in his iconography. Thus, the diadem of feathers is interpreted as a crown in his two engraved portraits of the Brazilian king, a woodcut in the *Cosmographie universelle* and a copper-plate in the *Vrais Pourtraits*.[85] In the same way, the ring of feathers worn around the haunches, the bead necklaces, the crescent of pectoral bone and the pendant earrings – which one might compare, incidentally, to those worn by Henri III in the preliminaries to his prosopography – all combine together to help define the pomp and magnificence of a truly royal 'costume'.

Now we know from ethnologists' accounts that such attributes were not the exclusive preserve of chiefs, but were shared among the whole class of warriors. The only object that might perhaps have fulfilled the function of insignia proper to a principal dignitary was the baton tipped with a tuft of feathers. But even this was less a mark of authority than an accessory to the dance, indispensable to the celebration of certain festivities, as Alfred Métraux has remarked.[86] Perhaps it was simply a question, in both cases, of a club; here schematized to the extreme and decorated with a collar of fine down, and used in combat – especially for the ritual slaying of a prisoner. But whatever its exact form and definition, it is clear that in Thevet's eyes and those of his illustrator this instrument fulfilled the office of a sceptre; which is why it is placed in the right hand of the 'demigiant'. The simple proximity, in Thevet's gallery of *Vrais Pourtraits*, of this exotic figure with the *imperator* Caesar ('Julius Caesar, First Emperor': VIII, 135), with Suleiman the Magnificent (VIII, 146), or with Charlemagne (IV, 3) shows a quite remarkable adaptation of the 'ethnographic panoply' of Brazil to the universal, timeless mode of royal representation. By a resemantization of his dress (reproduced, however, quite faithfully), King Quoniambec accedes to a sort of monarchic eternity.

Marc Lescarbot, in his recourse to Jean de Meung, observed certain distinctions according to regions and epochs, and fixed narrow and exacting limits to the prerogatives of the American king – who, he notes, had that status only for the duration of a war or expedition. Furthermore, he is said to govern, like the ancient German kings evoked by Tacitus, 'more by example than by command'.[87] On the other hand Thevet, far from similarly maintaining the prudence of a historian and the scruples of a comparatist, tends to perennialize the status of the Indian monarchy. Hence the 'summit conferences' to which Quoniambec is periodically invited by Villegagnon, where the two sovereigns – the French viceroy and the Tamoio king – debate in the presence of their counsellors ambitious designs of conquest: 'from time to time he was called by our chief to confer with him about discovering what there was of value and interest.'[88]

It is significant, too, that we see the myth of Quoniambec progressively gaining in precision with the passing years. By the time of the *Vrais Pourtraits* there is no longer anything risible about him; he deserves to figure fully in the ranks of *proceres*. Furthermore, there does not exist in Thevet's writings – by contrast with those of Jean de Léry or Théodore de Bry – a unique, homogeneous and transparent vision of the American Savage (who on the contrary is dislocated between insults about his laziness, stupidity or incessant larceny, and dithyrambs eulogizing his bravery or hospitality); but his image of the chief is there to make up for this patent lack of symbolic cohesion. In the cosmographer's abundant work Quoniambec is one of the rare cases in which the American acquires true individuality, and thereby escapes from that litany of discordant stereotypes.

In other words, relations with the Other are posited by Thevet only on the political level of command. The 'king' of Brazil – not the basic Indian, one might say – is considered by him as the only valid interlocutor; a channel opened up at the price of the slight 'fiction' that the American monarchy represents. Thus, if there is not, with the 'cosmographer to four kings', a general type of 'noble savage' analogous to that found in Léry's *Histoire* or Montaigne's *Essais*, there none the less exists in his work a political paradigm; it goes by the name of Quoniambec.

One can imagine the misunderstandings to which the flatterer of some chief of a Tamoio confederation exposed himself. Léry's riposte began the year after the *Vrais Pourtraits* appeared, in the third edition of his Brazilian *Histoire* (1585). He ridicules the 'scientific Quoniambec',[89] spangled, as if he were a bed of flowers, with improbable virtues. His 'palace' is reduced to a 'pigsty', and his vast 'territory of obedience' to a few acres of brush and forest.[90] From a strictly ethnographic point of view, there can be no question that Léry was right. We know today that the Tupinamba chief's authority did not have the exaggerated reach that Thevet supposed. A war leader above all (as was already correctly interpreted by Marc Lescarbot),[91] he was chosen for his courage and the number of prisoners he had captured and immolated in a communal banquet. This power was, therefore, rarely hereditary. Beyond the military marches where his authority seems to have been effective, the power of the 'king' fell away in times of peace, within a social organization that was of the loosest kind; the warrior returned to the *maloca* and to an extended family under the authority of a venerable elder.

It would even be easy to oppose to André Thevet his own affirmations elsewhere (in the *Histoire de deux voyages*, for example), that the rights of the chief amounted – temporarily, however – to so many duties. In accordance with a schema that the ethnologist Pierre Clastres has recently, on the basis of the neighbouring case of the Guayaki of Paraguay, theorized on the (somewhat ideal) model of 'society versus the state',[92] the chief is defined by his ability to ensure provisions of food and to meet the needs of the group. He therefore seems to be in debt to the community as a whole. Exchange is to his detriment: if he commands, during the brief periods of seasonal war, that is, in August at the time of fishing for a certain type of fish, or in November when the maize is ready for the making of cahouin, he must in return ensure the provisioning of his men and sacrifice his reserves for them: for 'The rest of these poor people go to their villages in order to be fed for a time at the expense of the entrepreneurs.'[93]

The chieftainship – and this was particularly true of the Tupinamba – thus appeared as an intermittent phenomenon, which was expressed moreover in obligations of an economic and military order (haranguing and feeding the troops, and leading them into battle); rather than in any real, durable prerogatives.[94] In this regard, Pierre Clastres has shown that the triple privileges of the chief (also recognized by Staden, Thevet and Léry) – rights in public speaking, wealth in food, and polygamy – in practice constrained him to a continual linguistic, economic and sexual expenditure, whose aim was to prove at any given moment, by a constant wasting of his forces, the 'innocence' of his power.[95]

On this point, Léry was a better ethnologist than France's cosmographer royal. However, he perhaps failed to see the essential point of his adversary's proposition. To valorize, against all likelihood, the person and function of a 'king of America' amounted to envisaging as possible a colonial enterprise of vast scope. The homology of social structures on both sides, the similar conceptions of seigneurial dignity and the 'eminence' recognized in this 'principal' of the savages were made to order for facilitating a peaceable conquest. The Indian monarchy was a myth indispensable to the establishment of relations of alliance with the new peoples, and then of gaining jurisdiction over their territory.

It seems, however, that the myth of the Indian monarchy surpasses in its symbolic implications a strict utilitarianism aiming towards colonial ends. With Thevet, the political use of fiction does not at all exclude wider resonances that make the latter contribute to a global and period-oriented vision of the history of peoples. We recall that the chapter of the *Vrais Pourtraits* devoted to Quoniambec opens with a parallel between the Old World and the New, which proves more favourable to the latter. If America appeared, in certain aspects, less governed than her European and Asian cousins, she surpassed them on the other hand, not only in the 'amenity and fecundity of the land' but also in the 'graces' with which her inhabitants were endowed. In other words – those Montaigne would use a little later on – all that this infant world needed was a pedagogue in order to realize qualities that were as yet in a virtual state, but which could in time make the pupil rival the master.

Thevet, whose expansionist optimism corresponded rather well with that of the conquistadors, would none the less have approved of their opponent Montaigne when the latter prophesied, 'that other world will only be emerging into light when ours is leaving it. The world will be struck with the palsy: one of its limbs will be paralysed while the other is fully vigorous.'[96]

The 'peerless amazement'[97] that gripped the cosmographer as he observed the new horizons is inseparable from the perspective of a crisis and a regeneration of the world. According to the thesis of a *translatio imperii* (transfer of power), moving from east to west following with the daily course of the sun, America represented the future for a torn and moribund Christianity in Europe. In such a context, the theme of gigantism attested to the almost intact youth of a continent for which the process of decay had barely begun. Whereas the giants of the Old World were from a time before the Flood and no longer existed except – witness the corpse exhumed at Saint-Germain-des-Prés – in the form of scattered bones,[98] those of the New World were indeed still living.

One can therefore see that a figure such as Quoniambec could come to represent the differential march of history. Again, the contrast becomes all the more striking, in the Rouen interview described by Montaigne in relation to the subject of 'cannibals', between a vigorous Brazilian captain and a fragile crowned child who ruled over France. The movement that involved a transfer of empire – of political power and of natural vitality – from one side to the other of the ocean thus appeared irreversible.

However, it is worth refining this perspective somewhat: for American gigantism was coloured by a fundamental ambiguity. If it appears in fact, in our time, as a vestige of an earlier stage of humanity, this can be interpreted in two totally antithetical senses. One might either insist – as was the case with Montaigne – on the future possibilities with which that infant world was pregnant; or one might emphasize the archaic (and as such, monstrous) nature of those peoples who had hitherto escaped the common revolution of the world, and whose excessive size manifestly contradicted the universal 'ageing' to which the sense of history led. In a theocentric vision of the world giants represented, indeed, a diabolical anomaly. How, from Pigafetta to Thevet, could one explain – other than as a manifest exception to the Flood – the survival of those titanesque creatures, the Patagonians?[99]

There was scarcely any doubt, at the time, that the athletes encountered by Magellan on southern shores were related to the criminal generations that had gone before and caused, by their excesses, the 'universal inundation'. Gigantism then became the negative sign by which the superabundance of sin and blasphemous pride assumed an almost concrete dimension. Or, to follow Thevet's reasoning, it was because they showed an exaggerated confidence in their physical strength, in the matter with which they were engorged, that the giants of the antipodes could 'hold in contempt the celestial spirits, eat and devour men, and in general give themselves over to all such impieties as those who lived before the Flood wallowed in.'[100]

How might it be possible to reconcile these contradictory myths coexisting, for example, in the work of the cosmographer? There one finds, almost side by side, the laudative portrait of the 'great king' Quoniambec and a frankly pejorative depiction of Patagonian 'Giantery'.[101] In fact, the difference in the treatments reserved for the equally imaginary) figures of the giants of Brazil and Patagonia depends to a large extent on the relationship of communication in which they are situated. The gigantic in its Patagonian avatar subsisted, as such, through a lack of relations with the outside world – in the event, with the conquerors from Europe. It was a negative gigantism, because it was born of exile and nurtured by an age-old separation. The 'demi-giant' Quoniambec, on the contrary, resulted from the contamination of a Brazilian ethnographic reality by the image of the feudal prince imported from the Old World. Whereas the Patagonian was the hypertrophied fruit of an isolation, the Brazilian appeared as the result of a conjunction. His lesser degree of 'procerity' – he was at most a 'demi-Patagonian' – indicates the hybrid nature of that mythic composition. As an intermediary between demoniacal titans who had escaped the Flood and the puny humanity of a world in decline, Quoniambec was an ideal middle term by which to join up those two halves of the universe and re-establish, between man's future and his past, the necessary unity and continuity.

The opposition between the anarchistic Patagonian and the monarchic Brazilian could thus be explained in terms of spatial relations. The withdrawal of those monsters of iniquity to the furthest confines of the inhabited earth could, on its own, account for the survival of the Patagonians down to the dawn of modern times. Or, following the immanentist hypothesis formulated by Girolamo Cardano in his treatise *De rerum varietate* in 1557, the gigantism of those distant peoples had resulted from autochthony and autarky: 'It was because they lived in the same country, with the same customs, without travelling nor receiving travellers; so that all their qualities, good and bad, were developed to the extreme degree.'[102]

The cold climate of Patagonia also favoured the propensity of its inhabitants to grow, and the absence of openings on to the rest of the world aggravated that natural disposition. On the other hand, the great discoveries, which created new circuits of exchange and opened up, one fine morning in the winter of 1520, the plateaux of Patagonia, would have for their consequence the progressive domestication of those disparities born of autarky. Maritime travel compromised in every possible way the gigantism of the Americans: it was threatened, from the genetic point of view, by sexual, alimentary and microbic exchange; and on the level of language it was ruined, as a legend generated by distance and the difference of languages, from the moment relations of communication were established.

Significantly, however, the various versions of Magellan's 'invention' (discovery) of the Patagonians all emphasize the impossibility or failure of the exchanges attempted by the navigators. At the end of a brief interview paid for, on either side, with volleys of arrows and cannon fire, the gigantism of the barbarian remained intact in its isolation. The Spaniards went back aboard their ships and left those inhospitable shores, where there would arise, barely perceptible among the mists of a dream-world, the silhouette of the fearsome Patagonian.[103]

With singular prescience, Girolamo Cardano discerned in the generalized exchange spreading over the entire surface of the globe the principle of a triumphant *entropy* – the same one whose deleterious and uniformizing effects would be deplored by Claude Lévi-Strauss four centuries later.[104] But whereas the author of *Tristes Tropiques* sees in it an ineluctable impoverishment of the field of research promised to the anthropologist (who is thenceforward reduced to the unenviable role of *entropologist*), Renaissance men like Cardano could only rejoice in this reduction of differences, this correlative simplification of the world. Until its slow work of erosion of extremes was completed – and it would take three hundred years for the fabulous 'procerity' of the Patagonian to be washed out – the generation of the Bolognese doctor was able to contemplate in its profuse diversity the entire order of the universe, at last rendered visible.

It was quite a different matter with the myth of the demi-giant Quoniambec; who was the son, and not the victim, of the growing entropy. As both the emblem and the product of an encounter, he played the role of collaborator, in both the economic and the political senses of the word, since he did not hesitate to give to the conquerors the strategic keys to his vast realm. It was he, notes Thevet, 'who advised us to take control of the rivers and islands around us, and build forts there for our defence'.[105] The active role he played in the joining-up of the two extremities of the world no longer needs to be demonstrated, and his physical stature was augmented by the prestige conferred on him by his dialogue with the West.

Thus we find that the monarchical principle had, for our titular cosmographer, a constant reference. Having for his object the universal description of the world, and cast as such in the role of the courtier making to his king an offering of the innumerable treasures of Creation, Thevet construed relations between nations on the model of interindividual exchanges between monarchs. Whence his 'invention' of Quoniambec, who reflected back to the 'Prince of Europe', in a plausible anamorphosis, his own image travestied into that of an Amerindian Hercules. In the interval between these twin figures, the logic of gift and counter-gift could freely play – in the direction of an expansion of the power of the French monarch, who was destined in the more or less long term to encompass his transatlantic double.

Now this triumphal myth would soon be ruined by the polemic that was rekindled against the cosmographer of four kings by Jean de Léry and relayed by the diligent pens of historians like Urbain Chauveton or Lancelot Voisin de La Popelinière. There can be no doubt that Léry and his fellow Protestants harboured a certain enduring resentment against the emanation, in whatever form, of an authoritarian and restrictive power from a single person. Bearing witness to this are the long debates reported in chapter 6 of Léry's *Histoire d'un voyage faict en Brésil*, where the point appears to have been as political as it was theological in nature. To the 'tyrant' Villegagnon, who was a rather tepid Catholic (but one concerned to be sole master on board, and who ended up sending three restive Huguenots to the bottom of the bay of Rio), were opposed the fourteen 'Genevans' who, for their part, favoured a collegiate administration for the colony of Refuge. Opposite a monarchy decreed to be illegitimate, since it had broken with the Gospel, was sketched out the formula for a consistory.[106] The quarrel quickly took a deadly turn: refusing to be 'subjects' of the king of Antarctic France,[107] Jean de Léry and his companions passed into open rebellion, and rejoiced to shortly afterwards see the flesh and 'great shoulders' of the Knight of Malta [Villegagnon] 'serve as food for the fish'.[108]

This hatred of tyrannical monarchy is to be found in his caricature, to say the least insistent, of Thevet's Quoniambec. Clearly, it was not on a whim that Léry took as his favourite whipping-boy the 'demigiant' of Ubatuba, a worthy counterpart to Villegagnon in Cyclopean stature. Quoniambec incarnated, in Léry's eyes, an unacceptable concept of native society, for which the pastor long entertained a nostalgia and which represented for him, despite the ruses of Satan that were to be found everywhere, a last corner of Eden. In the Brazilian utopia of his *Histoire*, there was no place for an Indian monarch.

Léry repeatedly places the accent on the role of the elders, who in interminable harangues exhorted their folk to exact vengeance[109] and led the tribe when it made its migrations to another gardening area or into enemy territory.[110] In so doing, he deliberately confuses two instances that Thevet had taken care to distinguish: the military chief, who according to the cosmographer was a man in the full vigour of his prime; and the 'old man' who to an extent played the role of the sage, and whose great family assembled in the *maloca* to listen to his bellicose speeches.

This confusion allowed Jean de Léry to eliminate not only the too cumbersome person of Quoniambec, but also the specific function he fulfilled in the Tamoio microcosm as Thevet had represented it. Léry did not miss a single chance to recall the fundamental inexistence of that directorial instance. There was no 'field marshal' nor 'any other who in general matters rules over the houses',[111] he noted, apropos of the long marches that, 'without orders, yet without confusion', carried the whole group – men armed with bow and club, women bent under the weight of portage baskets and children – into the field of combat. From this image was constructed a social and military utopia in which the community, although free of the orders of a chief, presented an admirable cohesion and executed manoeuvres with perfect co-ordination. Only an oxymoron could express this miraculous state of 'non-anarchic anarchy'.

The society of the 'Toüoupinambaoults', if we are to believe Léry, rested on an egalitarian principle, such that there were among them 'neither king nor princes' and that all the warriors were 'each almost as much a lord as the others'.[112] He could hardly have opposed a more flagrant denial to the Thevetian fiction of the 'great king' Quoniambec.

In Jean de Léry's hypothesis, the myth of the Indian monarchy had arisen from a misunderstanding. It was a simple error of translation that, from the outset, had generated the fantastic idea of an Indian monarch. A dialogical exchange in his *Colloque de l'entrée ou arrivée en la terre du Brésil... en langage sauvage et en François*[113] proposed substituting, for the 'frustratory' term 'kings', that of 'heads of families'. It is this same expression that again appears in Léry's work in contexts of welcoming and hospitality, where the 'Elder' opens to the stranger his house, offering him food and drink, and sometimes even his daughter.[114] The *Moussacat*, in his function of a generous and attentive host – and we know that this figure would have a long posterity, at least down to Diderot's Tahitian – has replaced the fearsome *Morbicha* dear to Thevet, whose thundering authority (as a sort of 'cannibal Jupiter'), far from being restricted to his immediate family, extended over an entire people.

If the 'head of the family' finally supplants the chief, it is because the family cell has, from Thevet to Léry, effaced the state. For the monarchical system, incarnated and magnified by the person of the gigantic King Quoniambec, was substituted the intimate domestic circle grouping around the father, a mother and a child. An eloquent image of this spatial and sociological restriction is offered by the first engraving encountered in leafing through Léry's *Histoire d'un voyage*, which might be dubbed 'Brazilian family with pineapple'.[115] Tenderly laced around a naked warrior armed with bow and arrow is a Tupinamba woman, who is placed slightly in the background and carries a plump, chubby baby in a traditional sling. The pineapple in the foreground, the fruits of the 'choine' tree in a cupola, and the hammock in the background all confer on this idyllic genre scene the indispensable touch of exoticism.

An Edenic image (and one just as fictive, after all, as the *terribilità* of the cannibal king), this woodcut bears witness to a decidedly more modern vision of the American, the aftermath of which has not yet ceased to haunt the oneiric space mapped out by modern anthropology in the margins of its field of investigation. No more solid proof could be needed than the favour enjoyed, today as never before, by Léry's *Histoire*. Do we not see a direct extension of his iconography in the extraordinary photograph that the author of *Tristes Tropiques* entitles, precisely, 'Intimacy', and which shows, grouped together in smiling complicity and contentment, the Namikwara family?[116]

Léry eclipsed Thevet, in the same way that the Brazilian family has sapped the prestige the cosmographer sought to confer on his improbable 'demi-giant'. However, in the occultation of the political aspect that took place correlatively, perceptible already in Léry and taken up again by modern ethnologists,[117] it might be appropriate to see the stigmata of a guilty conscience in the face of a New World that, in the decade 1555–65 (and for a while at least afterwards), with France's repeated setbacks in America, came to be seen as something not so much to be conquered as to be dreamed about, in the ineffable mode of nostalgia.

5
Cartographics: An Experience of the World and an Experiment on the World

---

Writing has nothing to do with signifying, but everything to do with surveying, cartography, and even with worlds to come.

Gilles Deleuze and Félix Guattari, *Rhizome*

The unity of cosmography: the case of Guillaume Le Testu

Towards the middle of the sixteenth century, the word 'cosmography' acquired an incontestable vogue: one finds it among technicians of nautical science, who were 'unlettered' and expressed themselves in fractured French, such as the Portuguese Jean Alfonse (of Saintonge, as he was called)[1] as well as among learned renovators of ancient geographical science such as Waldseemüller, Peter Apian or Sebastian Münster.[2] The discredit into which the concept would later fall perhaps had to do partly with the fact that it was not socially discriminating: it threw together pell-mell, placing them on the same level of prestige, the unschooled pilot or 'mariner' who could steer by dead reckoning, draw portulans and sometimes even make maps, and the genuine savants in their studies, who worked on documents, checked new reports against the treasures in their libraries, and tried to renovate the compilations of ancient geographers.

A characteristic usage of the term, in this respect, was that of the Le Havre pilot Guillaume Le Testu in the dedicatory epistle of his *Cosmographie universelle*, an atlas painted on vellum paper and finished on 'the fifth day of April 1555 before Easter'.[3] He borrowed, in addressing Admiral Gaspard de Coligny, a large extract from Thevet's dedication of the *Cosmographie de Levant* to François III, comte de La Rochefoucauld. The passage is introduced by the following rhetorical question: 'Who is this new Cosmographer who, following on the heels of many highly renowned authors, both ancient and modern, has undertaken to invent new things?'[4]

By contrast with his predecessor, Le Testu limits himself by suppressing the pedantic reference to the legendary Anacharsis, the Scythian whose long voyages left him wiser than the most philosophical of the Greeks.[5] But we can already see that in the course of this pure and simple take-over, the word 'cosmographer' has changed its meaning. Being common to two distinct professions, that of the wandering Franciscan and that of a captain expert in the nautical arts, it passed without difficulty from one personal style to the other.

The rhetorical response, of a topical nature, was also borrowed word for word from Thevet, who optimistically evoked the immense nature recreated from age to age by a generous Providence whose actions were always one step ahead of the timid efforts of human knowledge. The eulogy of cosmography was thus transmuted into a hymn to *Natura naturans*, and to the expansion of knowledge about the world: But I would reply to them [who would ask such a question], that Nature was not so constrained or subjected by the writings of the ancients that she could have lost the power and virtue of producing new and strange things, beyond the things of which they wrote.'[6]

Even if the slight dig Le Testu aims at the ancients for lacking diligence and curiosity recalls the tone of his predecessor from Angoulême, it is undeniable that the topos seems more appropriate at the opening of Le Testu's oceanic atlas than at that of the Levantine compilation of the Franciscan friar.

Before the division of geographical labour that would become the rule from the time of Ortelius and the Antwerp school, at the precise moment when cosmography was becoming an outmoded concept, the latter supposed at the same time a mathematical basis and a technical instrumentality. Just as the term 'portolan' covered two distinct and complementary objects (the portolan map and the portolan rutter, or portolan in the strict sense), one could say that cosmography doubled itself into twin forms of cartography – the *mappa mundi*, or the atlas that cut it up into sections; and the book that was a collection of histories framed by the fourfold rubric of the four continents. In this sense, as Michel Mollat has observed,[7] the couple formed by the *mappa mundi* and cosmography was the heir to a homologous and anterior couple constituted by the portolan map and the portolan rutter. This new dyad was itself embraced globally, under the name of cosmography.

Thus Thevet, like Münster or Postel before him, considered himself as much a cartographer as a geographer. And his maps – notably that of France, now lost – are the part of his work most consistently and extensively cited. Their use-value, it seems, only slowly decayed, whereas the 'literary' side of his cosmography very soon ceased to be read. If Thevet was still remembered in the classical age, it was on account of the maps and plans bearing his name. The most illustrious Cardinal Francesco Barberini and, later still, the royal librarian Jean-Baptiste Bourguignon d'Anville did not disdain to seek out and even copy the dispersed plates of the unpublished *Grand Insulaire*.[8]

We know, furthermore, that Thevet's undertaking ended – or rather, failed to – in an atlas: that, precisely, of the unfinished and soon dismembered *Insulaire* that was without doubt the most original of his productions. Begun about fifteen years before his death, it was announced in several chapters of the *Vrais Pourtraits* of 1584.[9] It was based on the Venetian *isolarii* of Benedetto Bordone and Tommaso Porcacchi da Castiglione, and pushed the principle of an encyclopaedia as an archipelago to the limits of the possible. Some 200–350 maps of islands, 'both inhabited and uninhabited', filled it; of which a good third survive, covering in almost equal measure the ocean (vol. 1) and the Mediterranean (vol. 2). Or, to retain the image of the world dear to Thevet, the set is distributed between the oceanic Mediterranean, recently discovered and reconnoitred by modern navigators,[10] and the ancient interior sea to which the nautical audacities of the Greeks and Romans were effectively limited.

Now the *Grand Insulaire* was also a rutter. Thevet, who nourished its substance with maritime pilot books borrowed from Alexander Lindsay (via Nicolas de Nicolay), as well as from Pigafetta and Jean Alfonse de Saintonge,[11] called his last work *Le Grand Insulaire et Pilotage*, a great atlas and pilot-book of the world's islands. By that fact, the 'cosmographia bifrons' evoked above joined up, at the end of its evolution in Thevet, with its dual ancestry: the *portolano–compasso* couple. But what appears to be a return to the origins of Renaissance cartography should be considered, rather, as the completion of a cycle. In effect, an *isolario* or atlas composed exclusively of islands resolved in its way – the most elementary possible – one of the major difficulties posed by the cartographical construction of the globe at the time. The fragmentation and dichotomization it authorized provided a way of overcoming the hiatus that existed between a science of mathematical projection and the art of placing fragmentary empirical data on a canvas. That montage would, in a way, be left to the discretion of the user or reader.

But one difficulty was thereby obviated only by falling into another. Without an overall map, a guide to the map's orientation (a function that could not be fulfilled by the very schematic northern and southern 'half-parts' of the world placed at the opening of the manuscript), one fails to see how the practical mariner could make use of such a fragmented atlas, made up of indefinitely divisible and disjointed pieces.

Similarly, by scattering over dispersed chapters complete itinerary sequences, successions of capes, reefs, positions or soundings that he had taken from the works of Lindsay, Pierre de Garcie, Michel Coignet or Jean Alfonse, Thevet interrupted the practical continuity of navigation, isolating from it vestiges that were scattered among so many island receptacles, closed on to themselves. Thus exploded and cut up, Scotland, the peninsula of Brittany or the great Atlantic archipelagos share their drifting fragments with each other in the universal brew of this disordered maritime encyclopaedia.

This insularization of the nautical atlas can be seen as well in the dozens of marginal annotations Thevet made in the manuals of pilotage in his possession. In the *Grant Routtier* of Pierre Garcie, for example, the schema showing an elevation of the coasts of 'Lisle Dieux' (the Ile d'Yeu, in Brittany) is accompanied by a 'Nota'; while the rubric devoted to 'Groye' (Groix) is written off with the manuscript comment 'island'.[12] Brittany's 'Belle Isle' (fol. 55v) and the 'Isle of Oleron' (fol. 64v) are similarly the occasion for marginal 'headlines', references or reminders disposed here and there for the convenience of the copyist (for whom, no doubt, Thevet limited himself to preparing the task). In the same way, again, his copy of Jean Alfonse's *Voyages avantureux*, held in the Bibliothèque Nationale, repeats the comment 'isle(s)' opposite Alfonse's passages on the Island of Ferro or Hierro in the Canary archipelago, 'Terre-Neufve' [Newfoundland], Trinidad, the Islands of Hyeres (Hyères, off Toulon), Saint-Laurent (Madagascar), and Ceylon.[13]

A remarkable fact is that Thevet possessed a second copy of this work, in which he scribbled in an even more prolix manner, passing from wonderment to frank irony and even to the most bitter disdain. It is a veritable journal of his humours that these marginalia unfurl: 'a thing most false, says Thevet' (fol. 28v); 'not at all' (44v); 'Thevet says this is wrong' (46v); 'Halleluia, this is all wrong' (62r); 'you do not speak well, my good man' (63v). These less than delicate comments alternate with the word 'isle', repeated page after page: a rubric that drew from those texts the element essential to his own plans.[14] The manual of Captain Coignet, published (and therefore acquired) later, in which the cosmographer's interventions become more sparse, contains however the double mention of 'Isles des Asores' and 'Isles des Canaries'.[15]

The Bibliothèque Méjanes (in Aix-en-Provence) has Thevet's small copy of the book Pigafetta wrote, as both a voyage account and a guide to navigation. Here, the insular profusion is just as significant.

In this opuscule, printed in Paris around 1526 by Simon de Colines, the cosmographer has marked the progress of his reading with the lapidary mention 'isles', repeated more than twenty times in the margins.[16] Disobliging comments such as 'folly', 'this is false' or 'nothing here',[17] which had become something of a trade mark of the ageing cosmographer,[18] coexist with the gleaning of occurrences of islands, in this reading of the world as an archipelago. Peppering the established authorities with his punctual attacks, Thevet took from them descriptive islets: the fragments he considered useful for his own assemblage. Using a fairly exact image – say, the linear plot of Magellan's voyage – he retained only the punctuation. As we shall see, by playing with these natural articulations within geographical discourse the pilot Guillaume Le Testu had, well before Thevet, constructed the cosmographical fiction of a world without lacunae; one posited as perfectly knowable.

The imagination in the service of the map

We know the malevolent legends that circulated about Thevet: Jacques-Auguste de Thou denounced his impure use of 'road maps' and other books 'that are in the hands of the people'.[19] Peiresc ironized about the 'figures of islands' he made up 'by fancy', following the caprices of importunate visitors and mockers.[20] In fact, it seems that Thevet's practice in cartographical matters was situated, apart from a few deviations, within the received norms of the mid-sixteenth century. The cosmographer's art necessarily involved a recourse to disparate materials, often of humble extraction, and left the field open to the inventive genius of a manipulator.

A first necessity in such montage, as we have seen, was to graft the empirical on to the theoretical. To keep his atlas up to date, the cosmographer was led to integrate fragments of material without scale, positioned only by rhumb lines on the canvas of an armillary sphere, or on the grid of a geometrical projection of the sphere on to a plane surface. A second necessity was to articulate these pieces, all of diverse origin, on the preconstructed whole of a *mappa mundi*.

The first problem could be resolved by means of a very summary expedient: the simple succession of pages in an atlas. Thus, Guillaume Le Testu juxtaposes portolan charts to a series of projections at the beginning of his *Cosmographie universelle* (figures I to VI), without any scale of longitude. The various ways of constructing a globe – as a spherical triangle, on a boreal polar projection, on a symmetrical longitudinal canvas, in the form of a 'fully round glass bottle', or in a star projection with four equal spokes radiating from the Arctic circle – precede fifty map plates oriented only by wind roses, whose limits it is impossible to define since they include only a scale of latitudes and a scale of leagues for measuring distances.[21] It was left to users to join up the pieces of the puzzle and make them conform to one or other of the geometric models proposed at the outset, in whatever way they liked!

The layout offered by the *Isolario* of Tommaso Porcacchi da Castiglione, one of the models Thevet followed in his *Grand Insulaire*,[22] was rather similar. Here island maps, with no scale or outline of projection, are only summarily oriented by a wind rose situated ornamentally in one corner. In some of them (those of Taprobana and the Moluccas),[23] the 'equinoctial' or equator is drawn across the composition; or again, as in the map of Cuba, it is the Tropic of Cancer that runs along the north coast of the island.[24] To assemble the universal archipelago, the reader had at his disposal only two very schematic tables, pushed to the end of the volume. The first, entitled 'mappamondo', represents the outline of the sphere in symmetrical longitudinal projection. It is on this document alone that there figures a complete network of parallels and meridians.[25] The last map of the collection, a 'Carta da navigare' traversed by rhumb lines and graduated in latitude, was intended above all to be used by the practising navigator. A note instructing the good pilot in its usage shows him how to trace his route and hold his course. But it is unlikely that this planisphere of reduced scale, with its confused layers of winds, would have been much use to the seafaring man. It seems, rather, that this final universal map served Porcacchi as a technical alibi. By showing the amazed lay person the efficacy of the carto-graphic document, he gave *in extremis* (as a 'parting shot') the justi-fication for his collection. His appeal to the 'poeta buono' sidetracked the real destination of his atlas, while conferring on it the most emin-ent authority.[26] Thevet would behave no differently, in multiplying his sibylline asides addressed to the expert 'mariner' in the course of a heteroclite exposition whose only possible beneficiary was the bene-volent lay reader, the lover of rare and curious things.[27]

Already at the beginning of the fifteenth century, Christopher Buondelmonti had founded the fecund genre of the *isolario* and had been little concerned with precision or with providing the profes-sional user with reliable information. His *Liber insularum* was ad-dressed above all to a clientele of the sedentary, to whom was offered a voyage by proxy through the islands of the Archipelago (the Aegean Sea), as a pretext for retracing, in a sporadic and varied manner, the legends of classical mythology and the lost glories of a 2,000-year-old history.[28]

In his *Grand Insulaire et Pilotage* Thevet adopts a method in many ways similar to that of his predecessors, both immediate and distant. The three hundred or so maps of islands are preceded – or would have been, had the work been printed – by two maps of the northern and southern hemispheres in polar projection. Each of these 'half-parts' (*miparts*) of the world includes a scale of 'degrees of latitude' and another, spreading out fan-like from a median line, of 'degrees of longitude'.[29] The zodiac forms two semi-circles around them, from Taurus to Virgo in the northern hemisphere and from Gemini to Pisces in the 'mipart australle'. Eight chubby wind-cherubs complete the orientation of these halfling world-maps, and girdles of clouds fill the empty corners of the page. These two manuscript maps, placed at the beginning of the first volume, do not cover – far from it – the scope of the hundreds of island descriptions that follow; but they at least allow one to situate with some precision the main archipelagos of the ocean, such as the 'Isles espaignolles' or Greater Antilles, the Canaries, the 'Isles Moluques' distributed about the equatorial circle, or the 'Terres neufves', the mass of islands choking the St Lawrence estuary. If the 'Terre australe' of Portuguese origin, which here extends from the Straits of Magellan to the 'Sea of the Moluccas', reaches a tongue of land into the area of the 'Greater Java', the latter is no less insular in form: being separated only by an arm of the sea from 'Taprobane' or Sumatra. Also represented are the islands of Saint-Laurent (St Lawrence, or Madagascar), 'Nouvelle Guignée' (a New Guinea with unfinished coastlines to the south), and of 'Borne' or Borneo.

This cartographical framework in the form of a diptych is completed by two representations on a very small scale. These are, first, a map of the Atlantic at the top of the frontispiece, oriented with the north facing left and inscribed in the oval of a cartouche held up by two allegorical figures: Hercules (recognizable by his club and the lion skin on his head) and Atlas, who holds the celestial sphere in his left hand. These two personages are themselves standing at the prow and stern of a galley laden with nautical instruments.[30]

Also destined to take its place in the *Grand Insulaire* was the portrait by Thomas de Leu of André Thevet as 'insulist', with a forked beard. It too includes an element of general cartography. The left hand, holding the compass, rests on a terraqueous globe.[31] Outlined by Europe, Africa and America, all three mountainous and wooded, the Atlantic Mediterranean and its Pacific extension, with meticulously drawn waves, appear to be closed off to the south by a vast

D'André Theuet fut telle l'apparence
Qui le premier cheminant l'vniuers
Court Europe, Afrique, Asie immense
Premieres pars de ce Monde diuers
Et vid encor l'autre terre quembrasse
Le Ciel vouté sous l'Antarctique gond
Et le feisl voir ainsi qu'il se compaße
Descrit et peint dedans son Globe rond

T.D.L.fe   Antuerpiæ.

PLATE 1   Portrait of André Thevet the 'insulist', by Thomas de Leu of
Antwerp, c.1586. This portrait, which appears to have been intended for the
beginning of Thevet's *Grand Insulaire et Pilotage*, can be dated to the same year
as the copperplate maps engraved for the work. It shows, like the frontispiece
discovered by Marcel Destombes, a 'Mediterranean' Atlantic closed in by a large
'Southern Land'. Bibliothèque Nationale, Cabinet des Estampes, Rés. 7661
(Robert-Dumesnil 495.2). *Plate by courtesy of the Bibliothèque Nationale.*

La Geographie.   La Similitude dicelle.

La Chorographie de la particuliere defcription dung lieu.

Horographie (comme dict Vernere) laquelle auſſi eſt appellee Topographie, confydere ou regarde feulement aulcuns lieux ou places particuliers en foymefmes, fans auoir entre eulx quelque comparaifon, ou famblance auecq lenuironnement de la terre. Car elle demonftre toutes les chofes & a peu pres les moindres en iceulx lieux contenues, comme font villes, portz de mer, peuples, pays, cours des riuieres. & plufieurs aultres chofes famblables, comme edifices, maifons, tours, & aultres chofes famblables, Et la fin dicelle fera acomplie en faifant la fimilitude daulcuns lieux particuliers, comme fi vng painctre vouldroict contrefaire vng feul oyel, ou vne oreille.

La Chorographie.   La Similitude dicelle.

PLATE 2   Peter Apian, *La Cosmographie* (Antwerp: Gregoire Bonte, 1544), part I, ch. 1: 'Geography, Chorography, and a Comparison of them'. *Plate by courtesy of the Bibliothèque Nationale.*

PLATE 3 'In the seat of a ship, under the tutelage of the winds': *The Triumph of Magellan*, engraving by Hans Galle after a sketch by Hans Stradan, inserted into the *Americae Pars Quarta* of Theodore de Bry (Frankfurt, 1594; plate xv). *Plate by courtesy of the Bibliothèque Nationale.*

PLATE 4 Étienne Delaune, *Mêlée of Naked Warriors*, engraving no. 7 of the series *Combats et Triomphes*, before 1576, frieze on a dark background, 66 × 220 mm. Bibliothèque Nationale, Estampes (Robert-Dumesnil, IX, 287; André Linzeler, 281). This engraving, in which one recognizes by their rings of feathers and their characteristic clubs the Indians of Brazil, was inspired by Thevet's *Singularités de la France Antarctique*, fol. 71v, or his *Cosmographie universelle*, II, fol. 942v. *Plate by courtesy of the Bibliothèque Nationale.*

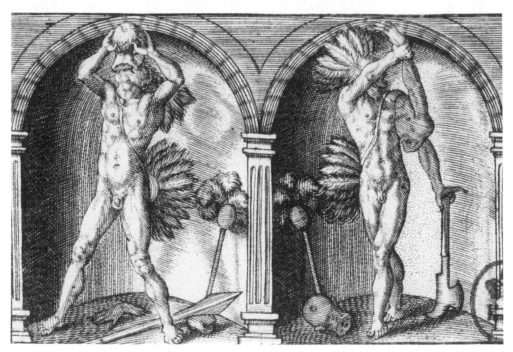

PLATE 5 Antoine Jacquard, *Les divers pourtraicts et figures faictes sus les meurs des habitans du Nouveau Monde* (Poitiers(?), c.1620), plate 6: *The Cannibal*, shown passing from the form of a naked man to that of a skeleton, by way of a flayed body. The maracas, clubs and axe were derived from various sources. *Plate by courtesy of the Bibliothèque Nationale.*

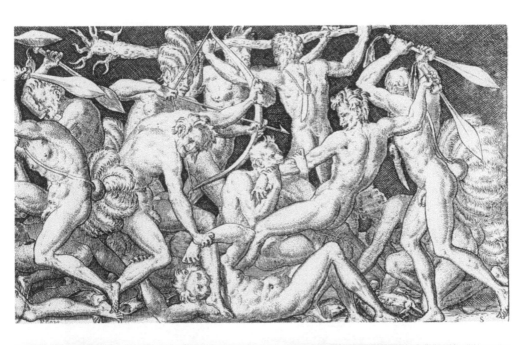

PLATE 8   *The Ruse of Quoniambech*, after André Thevet,
La Cosmographie universelle (Paris, 1575), II, fol. 952v, woodcut
(140 × 162 mm). *Plate by courtesy of the Bibliothèque Nationale.*

PLATE 9 *New Found Lands, or Isles of* Molues (cod),
copperplate (14.8 × 18.1 cm) from André Thevet, *Grand
Insulaire*. Bibliothèque Nationale, Ms fr. 15452. fol. 142 *bis.*
Plate by courtesy of the Bibliothèque Nationale.

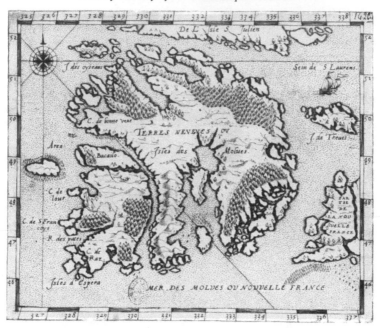

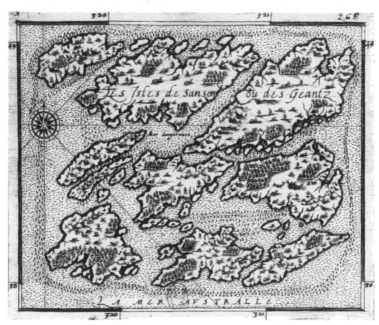

PLATE 10 *The Isles of Sanson, or of Giants,* no doubt
corresponding to the Falkland archipelago (being traversed
in fact by the fifty-second parallel). A copper-plate map
(14.9 × 18.1 cm) after André Thevet, *Grand Insulaire*
(Bibliothèque Nationale, Ms fr. 15452, fol. 268r). *Plate by
courtesy of the Bibliothèque Nationale.*

'Southern Land'. This configuration, already perceptible in the hydrographic chart of the frontispiece mentioned, accorded with the conceptions of Thevet – that friend of a Spain hostile to the 'continuous ocean' theory of the Englishmen Frobisher and Hakluyt.[32]

This cartographical tetralogy, situated at the opening of the work – the oceanic oval included in the frontispiece, the globe accompanying the bust of the cosmographer, and the twin maps of the hemispheres – did not constitute, it need hardly be said, an apparatus able to ensure the cohesion and practical efficacy of Thevet's insular atlas. The index of adjoining pages is incomplete and, behind the monumental and methodical aspect of the cosmographical façade just described, the sheer excess of his universal archipelago ruins any attempt to put it in order, even from the simple point of view of scale.

However, in each of his maps Thevet tried to forestall the risk of collapse and drift that was inherent in the project of an *isolario*. Mathematical cartography and practical cartography are not only juxtaposed in the density of the atlas, but are joined together at each moment of its progression.

Centred in general on an island, and more rarely on an archipelago, the maps of the *Insulaire* are small (15 × 18 cm) but reproduce the fragments of the universe on a large or medium scale. They are constructed, in the manner of portolans, around a network of rhumblines radiating from a wind rose; their intended usage being, as Thevet untiringly repeats, of a practical order. They were intended, far from the study geographer, for the 'good pilot' whom Porcacchi had earlier wooed. Consequently, the north–south lines of direction are not geographical meridians, but magnetic ones. These maps (sketched and engraved around 1586) represent, in this sense, an archaic state of the art, clearly falling short of the decisive innovation introduced by Mercator in 1569. In his celebrated planisphere on an isogonic cylindrical projection, loxodromic (rhumb-line) curves would become integrated into the system of the 'flat plate'.

But it is immediately evident that the maps of the *Grand Insulaire* combined two disparate modes of construction. By a collage, the cosmographer juxtaposed the space of the portolan, with its canvas defining areas of winds, with (in the border of the map, without any apparent relation to its *marteloire* or maritime background) a graduated double scale of latitude and longitude which, properly speaking, arose from the system of geographical projection. In the case of the two maps of Newfoundland, for example, one distinctly perceives the kind of rupture subsisting between a frame with lines uncurved from left to right – traces, it seems, of a polar conic projection – and the basic map itself, which was swept by rhumb-lines having no continuity

or liaison with the graduated canvas.[33] Furthermore, the straight lines
radiating from the wind rose appear to be in contradiction with the
clear curvature of the meridians and parallels, of which two brief
sections are drawn above and below, on the surroundings of the
document.

In effecting such a *bricolage*, the cosmographer forced together a
practical cartography based on the lore of sea-going mariners, and a
more theoretical cartography that subordinated the givens of experience to a rigorous method of geometrical construction. Almost realizing, thereby, the unitary knowledge of the earth that was still
problematical at the time, Thevet sought to be recognized as having
equal competence in the hitherto separate areas of theoretical cosmography and practical navigation. As was sufficiently well shown by
the double title of his island atlas, the two ambitions were for him
inseparable: on the one hand, to make a career as a savant and become
the equal of a Mercator or an Ortelius (whose approval we know he
sought);[34] on the other, to entertain a dialogue on an equal footing
with the most famous sea captains of his time, men like Jacques Cartier
or Jean Alfonse de Saintonge; to play, with the same assurance as
them, on the difficulties of nautical science, and to pretend to speak
their jargon.

The cartographic *bricolage* outlined here was not only the result
of a premature and clumsy synthesis between theory and practice; it
was also the effect of a general constraint that weighed on geography
at the dawn of the modern age – namely, that any given map was
never established on entirely fresh ground, but always inherited from
previous maps a not inconsiderable – even a preponderant – share of
its information. Even in the best of cases it integrated new data into
a received form or contour. The question of montage thus arose here
again; the risk of straying into the arbitrary was great, when it was a
matter of grafting the coastal outline of some newly discovered land
on to an ensemble that was already constituted and, furthermore,
'whole'. This gave rise to lapses, to necessary displacements of carto-
graphically existent parts, and to inevitable errors. The problem joined
that mentioned earlier: it was still, and always, necessary to ensure an
arbitrary liaison between a geometrically constructed whole, the form
of which was postulated as primary, and the parts that came along
afterwards to be integrated, with a rhythm varying according to the
progress of navigation and its divulgation.

In the end one is not surprised to see the triumph, in the late
and bastard avatar of cosmography that was the universal *isolario*,
of the art of *bricolage* in all its forms.[35] That spatial lability would
have, as its corollaries, cartographical anachronisms: thus Marco Polo

frequently coexists with Magellan and Jacques Cartier. The map did not reveal the state of the world at a given moment, but a mosaic of data whose chronology might extend over several centuries, the whole being assembled in a floating space. These driftings, at the same time spatial and temporal, conferred a dynamism and a prospective value on the map. On it were depicted not only lands actually known, but also those remaining to be discovered. This principle was generally applied by the Dieppe school of cartography, a worthy inheritor of the Portuguese tradition: for cosmography had a horror of the void.

A map at this time, although it is true that it had edges, could have gaps only if it masked them with a cartouche or with images of fabulous creatures. The portolan-map, of indefinite extension, joined the global *mappa mundi* in this graphic plenitude. Cosmography not only anticipated its own theorization by arbitrarily 'mounting' indications of reefs or capes within an empty structure; it fabricated whole territories in order to fill the empty spaces on its globe. It was in this way that Captain Guillaume Le Testu (after many others), his imagination perhaps stimulated by the *Chronique de Nuremberg* or *Schedelsche Weltchronik*, peopled the mythical depths of Asia and Africa with Blemmyae, Sciopods, Arimasps, and other such monsters. No doubt these *mirabilia*, which abounded in his *Cosmographie universelle* of 1556, descended in a direct line from those of medieval *mappae mundi*. At the beginning of the twelfth century, Canon Lambert of Saint-Omer, in the *mappa mundi* illustrating his *Liber Floridus*, had filled Asia – the traditional 'Land of Marvels' – with a fabulous bestiary borrowed both from the *Physiologus* and from Isidore of Seville. It ran from lions to elephants, by way of the phoenix (the symbol of a resurrected Christ). The most repulsive of such legendary creatures were cast out to the confines of the inhabited world – like the ferocious Amazons haunting the northern extremities of Scythia, or the serpents, dragons and 'cocodrilles' (crocodiles) that infested the torrid solitudes of Africa, that continent fertile in monsters and, as was well known, ever pregnant with novelties.[36]

For Le Testu, still – who in the meantime had claimed the further authority of Marco Polo the Venetian and that, more recent, of 'Emeric de Vespuce the Florentine'[37] – Africa contained snakes seven hundred feet long and capable of swallowing goats and bullocks; basilisks that could kill a man just by looking at him; satyrs, Blemmyae or 'men with no head'; Cynocephalics (dog-heads), and 'Colopeds'.[38] A battle between Pygmies and cranes is shunted on to 'India extra Gangem' (India beyond the Ganges); and Sciopods and Cynocephalics caper in the Himalayan foothills.[39]

But in this modern nautical *Cosmographie*, such marvels no longer altogether play the role assigned them in the theological and moral cartography of the Middle Ages. The world of Le Testu had lost its symbolic coherence and its geographical closure. But the fabulous creatures of the universal atlas then gained in picturesque value what they had lost in allegorical meaning. Beyond their evident ornamental quality, they paid tribute to the sacrosant principle of *varietas*. And they still designated, in a New France peopled with dog-headed or boar-headed men, or in 'Islands of Griffins' in the vicinity of the mythical Southern Land,[40] spaces that remained to be conquered.

This was particularly true of the oceanic region; that which primarily interested the mariner. Littered with '*bellues*' (whale-like monsters) spouting columns of water, traversed by ships bearing the royal fleur-de-lis or the arms of Admiral Coligny, and bordered by imaginary lands, the ocean furthermore harboured in its bosom an improbable austral continent, inhabited by the savages shown in ferocious confrontations or visited by unicorns. Le Testu, inspired here (as were his predecessors of the Dieppe school) by a Portuguese cartographical tradition,[41] makes no attempt to disguise the fictions that dominate his representation of the world. 'However, what I have noted and depicted is only by imagination', he declares in relation to this austral region, 'for there is no man who as yet has made a certain discovery of it.'[42] The key word here, 'imagination', forms a leit-motiv throughout Le Testu's commentaries; it announces the 'force of the imagination' later described by Montaigne,[43] and the power of which was such that it could create *ex nihilo* islands and empires.

In the dozen maps – out of a total of fifty-seven that make up the atlas – where he applies himself to describing the extent of this fictive geography, Le Testu in a way forestalls the progress to come in the area of nautical science. It was 'pending more certain knowledge' that he marked and named capes that were perhaps not at all the harbingers of vast and profound territories, but merely sporadic occurrences of volcanic reefs or coral atolls.[44] In this way, future explorers would 'be on their guard when, in their opinion, they were approaching the said land'.[45] Thus the fiction is prospective, yet corresponds to an immediate practical usage: it would spare the navigator the risk of shipwreck by prudently advancing into regions of the ocean that held the ever-present possibility of a sudden, abrupt and continuous coastline appearing in his path.

The heuristic role that devolved on to the cartographical imagination did not constitute, doubtless, any absolute novelty. Already Ptolemy had neatly situated the legendary island of Thule at 63 degrees of north latitude, half-way between the parallel of Rhodes and the North Pole. His 'invention', the sum of earlier and conflicting hypotheses, was intended to leave the door open for future research, while at the same time responding to a 'need for symmetry and simplification'.[46] As a border-island closing the European *oikoumene* off to the north, Thule had the additional merit of being situated at an equal distance, in longitude, from the meridian of origin in the Fortunate [Canary] Islands and the meridian of Rhodes and Alexandria, which cut the known world into two equal halves.

This double function – heuristic and taxonomic – is again found with the cosmographer Le Testu. He was perhaps indifferent to the spirit of geometry that had inspired the Alexandrian cosmographer, but just as desirous of arriving at a complete (if not exact) image of the terraqueous globe. The imaginary cartography he commended to the attention of his contemporaries allowed one, in the last analysis – as he repeatedly states in his Norman patois – to 'radresser les pieches' (orient the pieces) of the universal atlas.[47] Filling in the gaps of the *mappa mundi*, such insular or continental fictions – like the enlarged Java-la-Grande that extended towards the South Pole and joined up with Tierra del Fuego – suppressed any solution of continuity between the disparate fragments of an atlas composed of separate leaves. The 'joining-up' of patchwork pieces was facilitated, in the event, by the manifestly floating nature of these imaginary spaces: they could drift about at will and find their ideal placement at the junction of two maps whose scales did not exactly correspond or whose orientation left much to be desired. In one of the twelve tables of his atlas devoted to the Southern Land, Le Testu mentions an enigmatic 'cap de More' on its eastern side, but points out that it is 'only marked in order to align the leaves of this book'.[48] Thus, we see how cosmographic fiction played a hinge-like role within 'real' cartography – if, that is, it was possible to distinguish clearly between the one and the other. It was more than just a sort of makeshift filler of spaces, but a veritable structure in its own right, articulating the successive scenes of the description of the world.

At this stage of things, the difference between Ptolemy and distant inheritors of his legacy such as Le Testu and Thevet is striking. Le Testu's patchwork technique aimed not only at creating the 'right form' of the world – that *eusunopton* dear to the cartographers of antiquity.[49] Certainly, the notion of universal harmony had not deserted the language of geographers in the Renaissance;[50] but what mattered for practitioners like Le Testu was, more than the geometric perfection of the universal body, the fact that cracks were appearing all over the ancient representation of the cosmos. Cosmography gained a dynamism and mobility from the unstable borders that frayed out to an uncertain joining-up of the parts of the world, and that varied according to the whims of the map-maker and the progress of long-distance voyaging. Out of this play of earthly parts was born a moving space, as the cartographical imagination and a conquering pragmatism embraced each other tightly in the invention of new territories.

Fictions of new-found lands

Both constructive and projective, the cartography of the Renaissance was also political: for to speculate on the contours of an unknown land was to incite rulers to take possession of it. It was recalled above that the hypothesis of a southern continent, joining Tierra del Fuego and 'Java-la-Grande' to Antarctica, was of Portuguese origin. This particularly audacious cartographical extrapolation came to fill, as if by chance, the void of a theoretical possession: 'Java-la-Grande' was situated in the half of the world that belonged to the Portuguese, by virtue of the Tordesillas Treaty renegotiated at Saragossa in April 1529.[51] In annexing for his own ends this territorial invention, following his Dieppe colleagues, Le Testu pays homage to France and to his admiral, Gaspard de Coligny, to whom his *Cosmographie universelle* is dedicated. The ornamentation of the maps, notably by numerous scenes of naval battles in which the French always appear the victors, indicates unequivocally the political bearing that cartographical fiction had.

It even seems that the expected strategic profit served as a sort of touchstone for the fictions of the cartographer. In his 1566 *Mappemonde*, his last known work,[52] Guillaume Le Testu once again had recourse to his creative imagination, although with more restraint, it seems, than ten years earlier. He now drew a distinction between mendacious fiction, *la mensonge*, which he proscribed with horror from the 'true map', and what one might call 'fiction by provision' – an improbable toothing-stone that held the future in reserve. Faced with an unknown land, the cosmographer had a choice between two attitudes: invention, or pure and simple abstention. It was thus that, beyond a correctly represented Lower California, his coastline to the north-east runs out and, having not been 'duly' discovered, remains 'imperfect'.[53] Beyond Labrador is a similar vanishing outline. While awaiting fuller information – and these were, in fact, both questions that would engage geographical circles for many decades – Le Testu pronounces neither on the insularity of California nor on the possibility of a northwest passage.

But if in these cases he abstains from making the hypothetical delineation for fear – he says – of 'adding to this true map any lie', he shows less prudence with regard to the famous austral continent. Though more circumspect than in 1556, he indicates his sources and lets it be known that the Portuguese had fathered this cumbersome fiction.[54] While denying that the austral land has been discovered and declaring that he 'does not wish to attach any credence' to it, he none the less 'notes' it and deploys fabulous evidence for it in the whole lower part of his planisphere. Elsewhere too, in the vicinity of the Straits of Magellan (where the same continent ends, and where we see walking in single file a crocodile, a dromedary, a dragon and a panther), Le Testu inscribes this new commentary in the guise of an excuse:

> Magellan certainly passed through this strait and named it with his name and sailed along the coast of Peru; but there is no other man than him who has had knowledge of this coast. However, I have roughly depicted it here, following the advice of certain cosmographers but without wishing to affirm anything [about it].[55]

One notes in passing that for Le Testu, the outline on a map did not necessarily constitute an 'affirmation' of the real, and did not prejudge in any way the referent-space. In this sense, cartography was a process of experimentation; it made provision for the future, while orienting the present. One could, then, define it as an 'experiment on the world' in much the same way that *bricolage*, in Lévi-Strauss's formula, is an 'experiment on the object'.[56]

Such fiction was, by that fact, called to a plural vocation. Constructing the future, it had to offer possible alternatives. Thus we can understand, in Le Testu's *Cosmographie universelle* of 1556, the juxtaposition of two maps of Newfoundland with contradictory profiles: on the one hand, a promontory continuous with the North American continent, whose outline was inherited from a Lusitanian model;[57] on the other, an archipelago in the Dieppe tradition, recording in the form of a splitting-off the discovery and penetration of the Belle-Isle strait by Jacques Cartier in 1534.[58] Both models present complementary advantages from the point of view of tactical extrapolation: the continent offers to the discoverer and conqueror a resistance that remains, for the most part, foreign to the archipelago on account of the latter's fragmentary nature; but it is less 'deceptive' than the latter, which obliges the navigator to disregard it.[59]

In conclusion, one could say that the world according to Le Testu was like Marot's rondo: a structure at the same time closed and open.[60] Closed, because cohesion was maintained at the price of the most glaring fictions; and open, because these fictions were denounced as such: as papering over the cracks of a space in which something remained to be invented – an empty sea or a blank continent. The imaginary capes or floating archipelagos – the 'cap de More', 'île de la Joncade', or 'Ille des grandz hommes'[61] – that serve to position the successive leaves of the atlas would be a cartographical equivalent of the 'entry' into the rondo. Their return, from map to map, dissimulates a lack of meaning and foundation. It was up to the user – the reader of the poem, or the navigator at sea – to fill in that barely perceptible gap by restoring the implicit interval or trope.

Our examination of the paradigm constituted by the work of Le Testu may have allowed us to see, in his *Mappemonde* of 1566, a major step in the evolution it was undergoing. In the space of a decade we have passed from the *Cosmographie universelle*, the 'oriented' pieces of which map out the fiction of a closure, to an unfinished and open world where the cartographer, having constantly to interrupt its contours, gives up trying to show more than he knows. That evolution could only have taken place, one can surmise, with reluctance; whence the hybrid object that is Le Testu's planisphere of 1566. It is open to the north and closed to the south: casting improbable conjectures into the distant regions of the southern hemisphere while erasing, in the closer northern spaces, overly rash sallies of the forces of imagination.

That discrimination no doubt also corresponded to priorities of a strategic order. After the failure of the repeated efforts of Cartier and Roberval to discover the chimerical Saguenay and the expected route to the western ocean via Newfoundland and Canada, the northern confines of the New World were for a time left to the ambitions of the English (whom Thevet, as a worthy heir to the aims of the Dieppe school, tried to discredit). On the other hand, the myth of a southern continent would in France nourish, for another generation and beyond, dreams of empire and revenge. Proof of this was the treatise *Des trois mondes* by La Popelinière, which appeared in 1582 at the time of the war over the Portuguese succession.[62]

By comparison with Le Testu, Thevet for his part would remain at the first stage – that of his 1556 *Cosmographie universelle*. In his case, however, the anachronism was not as great as some have made it out to be. It involved not centuries, but one or two generations at most. The cartographical work that culminated in the *Grand Insulaire* was not that of a scholar from the Middle Ages who had somehow wandered into the Renaissance, but a 1550s cosmography that was out of step in the declining years of the century. Ending in the reign of Henri II, his work would have seemed anachronistic, then, neither in its method nor its means. But forty years later, the epistemological revolution brought about in the meantime by Mercator, Ortelius and their cohorts had relegated it to an earlier age. Thevet in his old age remained faithful to the system of representation of the Dieppe portolans, and ignored the innovations of the Flemish school. All it afforded him was documentary content – details of islands and towns, which he unabashedly plagiarized;[63] but its novel overall design escaped him completely.

Guillaume Le Testu's nautical fictions fulfilled conditions of technical instrumentality, at the same time as offering the French monarch and his minister, Admiral Coligny, the hyperbolic and anticipatory image of a maritime empire that was late in coming to fruition. In the maps of Thevet's *Grand Insulaire*, by contrast, the political function prevailed over practical usage and to some extent was incompatible with it. His representation of the 'Terres-Neufves' or 'Molues', for example, involves anamorphoses that defy any orientation or synchronization. The most manifest such alternation is a lateral inversion between east and west: his 'part of New France', which is none other than Cape Breton Island, separated from Nova Scotia by a narrow channel, appears at the right of the document instead of the left. But this unfortunate inversion seems to have resulted from no deliberate intention: Thevet's engraver, tricked by the technique of copper-plate engraving (which requires a reversed incision of the plate, in relation to the finished image), appears to have unwittingly produced a geographical monster.[64] That did not prevent him, however, from graduating correctly the scale of longitude from left to right, from 325 to 338 degrees (this being to the west of the reference-meridian passing through the Island of Ferro). As a result, the scale contradicts the outline.

On the other hand, the location of place-names was correct if one would only look at the map in a mirror. To the left, on the eastern side of the island (which should be on the right), one finds nomenclature inherited from Jacques Cartier or from even earlier than his voyage of 1534. Cape Raz (or Raze) terminates normally south-east of the peninsula of Avalon – here shown as an island separated from the main mass of 'Terres Neuves'. 'Cap de bonne veue' is that known today as Cape Bonavista, the name first given to it by John Cabot. The 'Isle des Oyseaux' (Isle of Birds) is Funk Island; 'Bacailo' is Baccalieu Island. 'Cap de lour' (Bear Cape) is one of the eastern capes of the island where Cartier, in 1534, saw a polar bear 'as big as a calf and as white as a swan'.[65]

A second anomaly on this map resulted, for its part, from a deliberate intervention on the part of the cosmographer. It is the presence on the right-hand side – that is, to the west – of an incongruous

'Thevet Island' at the latitude of Cape Breton. This could be the real Anticosti Island at the mouth of the St Lawrence, radically displaced towards Newfoundland; or it could be one of the Madeleine Islands, greatly magnified. The salient fact, however, it that at this late date (1586) Thevet could have felt free to rebaptize in his own name a land discovered more than half a century earlier. By an effect of megalomania (which shows up elsewhere in his work), Thevet pretended that he was the first European to set foot there. Yet it is an established fact that he never accompanied either Cartier or Roberval to Canada, and that he never even saw the shores of that great boreal island from the vessel that took him to Brazil in the winter of 1556.[66]

The function of such retouching is clear: it served as a signature. In confiscating for himself a place well known to pilots and geographers, Thevet claimed a sort of cartographical paternity; and indeed he had become, in these final years of the sixteenth century and well after the disappearance of the true 'inventors' of Laurentian New France, one of the best experts on these shores.

A twin sister of the preceding one, another 'Thevet Island' situated in the South Atlantic, is shown in the same atlas at the latitude of the possessions (by this time lost) of the Valois monarchy in southern Brazil. The ephemeral 'Antarctic France' of Rio de Janeiro, no less than the New France of the St Lawrence estuary, is marked with the seal of the king's cosmographer.[67] By means of such monogram-islands, drifting about at the whim of the signatory, the fantastic cartography of Thevet exercised a symbolic and purely theoretical empire over territories that in fact escaped the control of both the king and his geographer.

Such usages had, by around 1590, become rather unorthodox; one can scarcely imagine reputed savants like Ortelius putting their name in the natural cartouches formed by islands and islets on their planispheres. However, such a personal imprint attests, by its very exceptionality, to the confusion that always dogged Thevet, between the professions of the navigator (whose apprenticeship, as we know, was served 'in the seat of a ship, under the tutelage of the winds')[68] and of the savant in his study. Here again, his quackery had the function of unifying those two complementary but distinct tasks under the sole and sovereign responsibility of a royal cosmographer.

Political intentions, involving a not inconsiderable element of personal calculation, were expressed in an even more explicit fashion in his other maps relating to Antarctic France in the years 1555–60. The '*Gouffre* (gulf) of the river of Guanabara or Janaire' (Rio de Janeiro), which was also intended to feature in the *Grand Insulaire*, shows a shoreline redrawn in accordance with the strategic position held there

by the French for a few years.[69] In reality – but this was the cartographical image also found in his contemporaries, Le Testu, or Jacques
de Vaudeclaye of Dieppe[70] – the Bay of Rio de Janeiro does not have
the almost circular form that allows Thevet to inscribe it neatly within
the rectangle of his map. Further, the invention of a fictive 'Ville-
Henry' to the west of the entrance channel, at the approximate location where the Portuguese town of St Sebastian would be established,
seems to reflect the courtier's desire to offer his prince an image of
imperial glory radiating out to the antipodes.

The polemics engendered by this invention, beginning in the reign
of Henri II but regularly reiterated by Thevet down to the graphic
apotheosis of his unfinished *Insulaire*, reverberates from the works
of Pierre Richer and Nostradamus down to those of Jean de Léry,
Génébrard and Marc Lescarbot.[71] Its echoes bear witness to an almost
universal misunderstanding of the procedures of geographical writing
in the Renaissance. In creating out of nothing (and in a decidedly
inappropriate way, as history would soon show) an austral capital,
Thevet was simply applying the lesson he had learned from Guillaume
Le Testu. The very concept of an 'Antarctic France' that, with Thevet,
embraced the whole Brazilian region, represented right from the start
(with the publication of his *Singularitez* in 1557) an audacious prospective fiction.

A fragmented cosmos

Besides these creations that might be called original, and that show
Thevet as both a 'futurologist in the rustic state'[72] and a shrewd
observer of the geopolitical questions of his day, there are other maps
in his *oeuvre* that took up uncritically a legendary heritage that had
already long since been codified. Thus, half of the *Insulaire* was devoted to the islands of the Mediterranean; the essential object being
the Aegean archipelago, which is scrutinized right down to the mi-
nutest reef. In the tradition of predecessors like Buondelmonte,
Benedetto Bordone or Tommaso Porcacchi da Castiglione, Thevet
records the islet of Strongile, 'wreathed around with great cliffs rear-
ing into the air' and featuring a half-ruined temple of Bacchus in
which shepherds and flocks shelter in the heat of the day.[73] Or again,
the hollow island of Curco off Cilicia, whose deep caves gush with
running water;[74] and, above all, the reefs of the Caloyer which, being
steep and difficult to land on, sheltered Basilian Coenobites from the
aggression of the bloody barbarians.[75]

On other fronts, and in more distant geographical regions, Thevet amplified the givens of tradition, going so far as to propose three versions of the northern 'Isle of Demons' in the region of Labrador, where the unfortunate Marguerite de Roberval was supposed to have been marooned as punishment for her adulterous love affairs.[76] Going beyond the slender pretext of her 'tragic story' (which Marguerite de Navarre, and after her François de Belleforest, offered for the edification of Christian readers), the cosmographer enlarges this island frame to the dimensions of a triptych, and pours into it the substance of an exposition on demonology; as well as an incongruous refutation of the 'frozen voices' forged by 'Panurgic muck-rakers', in which it is easy to recognize the now-deceased Rabelais.[77]

It transpires, indeed, that Thevet censured the very cosmographical imagination to which he himself was generally indebted. The dual archipelago of Imaugle and Inébile, mentioned earlier and described at length in his *Cosmographie* of 1575, no longer pleased the insularist of 1586. Hence the absence of that archipelago structured by the difference of the sexes, and opposing a fertile island of Amazons to the sterility and ravages of an island of men.[78] Again, the 'Isles of Sanson [*sic*] or of Giants' situated in the South Atlantic might be nothing more than the re-emergence of a distant and fabulous folk tradition, contaminated by biblical memories. In fact this archipelago, striated by the oblique lines of a strait and lying at fifty degrees of south latitude, closely reproduces the configuration of the Malvinas or Falkland Islands, which are divided into two masses by the northeast–south-west diagonal of Falkland Sound.[79] The designation of these quite real lands by reference to legends of giants can be explained, then, by the proximity of Patagonia, visited in 1519–20 by Magellan's fleet and on the shore of which appeared, indistinct and menacing, Indians of tall stature clad in animal skins.[80] Furthermore, these 'Isles of Sanson' are attached to a whole chain comprising the 'Small Islands' (*Isles Menues*),[81] the 'Isle of Calis in the South Sea',[82] the 'Small Archipelago',[83] and the 'Isles of the Cape of Chile'.[84] These, extending from the Queen Adelaide archipelago to Tierra del Fuego, describe step by step the approaches to the hypothetical southern continent. This fragmentary and conjectural cartography echoes the inaugural voyage of Magellan, but also early Portuguese voyages in these parts, about which Thevet, no doubt through the intermediary of Jean Alfonse, seems to have been well informed.[85]

One sees from this example how the appearances of legend could cover or disguise, in Thevet's cartography, very real information. It was the same, proportionately, with the 'Isle of Demons' mentioned above. Playing on the heralds of the Counter-Reformation, Thevet multiplied at will the demons of Canada so that they might better be banished later, under the onslaught of French colonization. This spiritual 'cleansing' went together with the stripping of the forests. Since there would be little effort required, as Thevet assured, to clear these provinces and tear down the 'thick ramparts of forests' that provisionally prevented access to them,[86] the conversion of the natives should hardly pose any greater problem. Besides which, to turn the soil was already to strike at the heart of the reign of the *sauvagine* (the strange or 'fishy') over those uncultivated lands and inviolate waters. Wild animals flee the clearing where the colonist and priest set up their cabins; the devil retreats along with the woods. The 'Isle of Demons', as a fable turned to the purposes of a colonial plan acquired instrumental value: it was an inciting obstacle, and an indispensable counter-proof to the programme of polderization proposed for Anticosti and the principal islands of the St Lawrence estuary.[87]

Thevet's political fictions in the end ran up against a double limit, which their initial defeat sanctioned. It had to do on the one hand with their fragmentary nature, and on the other with their over-specialization. Works such as his *Histoire de deux voyages* or *Grand Insulaire* formed a literature for specialists, whose interminable asides would often alienate a lay reader or one not greatly versed in sea-faring matters. His insularist logic, everywhere in evidence in these two late texts, destabilizes all narrative, shatters all description, and saps the diegetical consistency of the hero-narrator, who is left in-capable of engendering any conviction. Thus, the section of the *Histoire* devoted to Canada juxtaposes, across three chapters, three lists with different natures and functions: an 'instantial' or 'recapitulative' list[88] with the alignment of rhumb-courses followed by the ship; a pro-grammatic list, which is the catalogue of the colony's expected riches; and finally a lexical list constituted by the 'little dictionary of the language of the Canadians'.[89] This model of paratactical, injunctive writing excludes any serious reading, but requires on the contrary that the lesson be broken up by action, and that a constant to-and-fro take place between the page of the book and real maritime space. In this, it recalls the pilotage manuals and nautical guides that Thevet, throughout his life, collected and covered with handwritten annota-tions. In plagiarizing Jean Alfonse, in glossing or 'insularizing' Pigafetta, Pierre Garcie or Michel Coignet, in pillaging Nicolay for his Hebrides or Orcades, Thevet always tried to maintain the unity of practice and theory: the central platform of a cosmography that was in decline.

But the attention he constantly paid to the practical aspect – one that he could only simulate, for want of being the expert pilot he was in his dreams – worked to the detriment of the universality that he none the less inscribed in the frontispiece to his works. The myopia of this landlubber, who could grasp the world only by strings of capes or dangling rhumb-lines, was reflected, as we have seen, in his cartographical fictions. His 'Java-la-Grande' crumbled definitively, and the southern continent (which Le Testu still showed as a solid mass extending from the Moluccas to Patagonia) gave way to a 'flotilla of little islands' so named, not because of their smallness, 'but because in any part there are so many of them that they are almost enclaved into each other'.[90] Thevet's fiction had become atomized by the time of the *Grand Insulaire* – the opposite of the continental welding, albeit precarious and approximative, that still presided over his *Cosmographie universelle* in 1556.

The consequence of this pulverization of the cosmos, which co-existed (somewhat contradictorily) with the hypothesis of an enclosed ocean,[91] was that, despite his cosmographical ambitions, Thevet most often limited his manipulations to the narrow horizons of topography or chorography. Was not his most famous and controversial fiction that Brazilian 'Henryville' whose crenellated ramparts barely rose above the depths of the tropical forest? At best, he redesigned the contours of a bay and rebaptized with his name an isolated mountain or two estuarine islands; but never did he broaden his remodelling out to the dimensions of a country or a continent.

There were at least two reasons for this. The first was the delay of technical means in coming to match his ambitions. In retaining the traditional and rather obsolete form of the *isolario*, the cosmographer could not pretend to be renewing the face of the world. In his desire to affirm at all costs the unity of practical knowledge and global science, he managed to retain of the cosmos only a loose and indefinitely fragmented matter, whose precise emblem is the dust of 'little islands' he gathered at the extreme margins of the *oikoumene*.

The second explanation of his failure has to do with the fact that, at the late date by which Thevet had decided to offer to the public a 'cosmographic body accomplished in all its parts',[92] manipulation on a very small scale was no longer possible, given the progress of geographical knowledge and the monopolization of the world by European imperialisms. The face of the earth was by now fixed in its main outlines, and the political balance that provisionally favoured a hegemonistic Spain denied our cosmographer, as a Catholic and Leaguer, any room for personal initiative in that domain. In the event, it would fall to the Protestant La Popelinière to reconstitute, as the basis for a projected empire, a vast austral 'third world' exhumed from Portuguese and Dieppe portolan lore.[93] All that remained for the dethroned cosmographer was detail on the very large scale, a domain where his 'imagination' could have free reign without impinging on that reserved for strategists and diplomats.

Thus he returned, almost without knowing it, to the origins of his work. The archipelagic world of the *Grand Insulaire* found its way, via a detour through cartography, back to the disparate magic of the *Cosmographie de Levant* or the *Singularitez de la France Antarctique*. The reign of profuse variety in the cosmos was in the end restored, in the swarming contours of the mosaic into which the universal atlas dispersed.

Epilogue: The End of Cosmography

It was in the decade 1550–60, at a time when the genre of cosmography reached its apogee, that Thevet found his vocation and realized his initial technical and documentary capital. From that time he built up his shipboard library (if we are to believe his ex-libris – ever subject to caution, because frequently antedated), which was based on Jean Alfonse, Pierre Garcie, Olivier Bisselin, Sebastian Münster and Manuel Alvares. It was around this time, too, that he gathered (again according to his ex-libris) exotic documents such as the *Codex Mendoza*, an Aztec manuscript from the first years of the Spanish conquest, and that he began, on returning from the Levant and then from Brazil, his collection of 'singularities': coins from the Greek and Latin peninsulas; feather ornaments from Mexico or Rio de Janeiro; coats, bolas and arrows that attested to the reality of the incredible Patagonians;[1] hippopotamus skins, toucan beaks, stuffed parrots and caimans. This corpus of memories relating to the New World would later be augmented, notably with materials from Florida; but it seems that those from Canada and Brazil, his favourite lands, were for the most part assembled by 1560.

However, in spite of these early touchstones, Thevet's cosmographical project only slowly revealed itself, gradually casting off the weight of polyhistory and emerging from a jumbled storehouse of *rariora*. And when, finally, public disfavour had changed him, the 'aged' cosmographer, abandoned by the humanists and by his scribes, found himself with a project that had lost its audience and its *raison d'être*. His late and futile rewriting of the *Grand Insulaire* as a *Description de plusieurs isles*, or his unfinished cosmetic treatment of the *Histoire de deux voyages*,[2] could not alter the fact that the cosmographic horizon of expectations had by this time already collapsed. It is significant that the public's resistance and that of the publishing industry, by then in a full state of crisis, manifested themselves at the time when Thevet was making his most personal efforts in his work. On the contrary, his works with the clearest success and the most durable audience appear to be his least original: the

Cosmographie de Levant and the collection of *Pourtraits* of famous men. The only works of his to be translated or re-edited in the following century, these did not contain any technical advances, were both without maps, and were addressed to a public with little education. One can even ask, particularly in the former case, whether Thevet had any effective control over the work of compilation that was done in his name by licensed and sedentary writers. What, precisely, was his own share in these distant avatars of a cosmography manifestly reduced to the status of a library inventory or a store of prosopographic archives?

In that 'method', which today seems surprising, there was in truth nothing exceptional. One might think, for example, of the two greatest prose writers of the French Renaissance, Rabelais and Montaigne. Had not Rabelais, whose forays into novel-writing were interwoven by many threads with Thevet's early cosmography, inserted at the end of his *Third Book*, modifying for the purpose their sense and compass, several pages from Pliny devoted to the names, habits and virtues of hemp, re-baptized 'pantagruelion'?[3] No one would dare reproach that 'doctor of medicine' for proceeding in such a summary manner to furnish an ending to his book. On the contrary, critics have endeavoured to interpret that literal borrowing by weighing up the parodic will behind such an appropriation, or by sounding the allegorical depths of a borrowing too obvious not to cover secret intentions. Thus the theft is metamorphosed into a hymn raised to the power of the human mind; the bland botanical and medicinal description drawn from the *Natural History* is enriched with unexpected bases at the same time as acceding, in this unforeseen avatar, to the 'highest sense' of pantagruelism.

One might apply the same hermeneutics to Thevet's wild usages, if the 'Pierre Ménard principle' – that of the apocryphal author of Cervantes' *Don Quixote* – really can be secondarily applied to all literature.[4] In two cases at least – that of the *Cosmographie de Levant*, which drew from the *Epitome* of Vadianus the spiritual sense of its progression; and that of the *Singularitez*, which order anthropological comparatism according to a model found in Polydore Vergil – a flaying of Thevet's text allows one to identify the true meaning and the profound coherence of what, from the outside, appears as a pure rhapsody or a cento with an itinerary for its pretext.

A comparison with Montaigne is even more instructive. The juridical gloss to which the writing of the *Essais* often comes close[5] – a structure of commentaries endlessly repeated and proliferating into indefinite growth[6] – corresponded to the numerous and insolent *marginalia* with which Thevet filled not only the books of his nautical library, but also the manuscripts of his own works that his more or less zealous scribes tirelessly copied. The most striking illustration of this procedure is the *Description de tout ce qui est comprins soubz le nom de Gaule*, a partial rough copy of the great *Cosmographie* in which Thevet multiplies stocks and grafts of his own invention.[7] In this way, he manages to ballast retrospectively the impersonal compositions of Münster, Nicole Gilles and Élie Vinet with his subjective experience. He could thus select the friends most necessary to his purpose, and rebel as he wished against the authorities his stupid, overly docile scribes recruited for him.[8]

The most remarkable characteristic in Thevet's case is the abundance of materials that have been preserved. Abandoned in full progress, his work finds its greatest richness in its very incompletion. Rare are the works from the Renaissance for which we today have four distinct drafts of the same page. But such is the case with whole chapters of the *Histoire de deux voyages* that are common to the *Grand Insulaire*, the *Description de plusieurs isles*, and the *Second Voyage*. As well, these different layers of manuscript, very close together in time, are often enriched or corrected reworkings of printed versions in the *Singularitez* and the *Cosmographie universelle*, dating from thirty and ten years earlier. The study or 'workshop' (*atelier*) in which was conceived this vast and controversial work can thus be reconstructed almost intact, offering a privileged field of investigation for the study of the genesis of a 'Renaissance text'.[9] Those bundles of anonymous labour that were enhanced page after page with ironical or bitter incidents give an essential glimpse into the way such second-generation writings were fabricated. The austere humanist erudition that had triumphed at the beginning of the century was disseminated to a wider public by the novel, picturesque means of a polemical irreverence that bounced from one interpolation to another.

In their very excess, Thevet's incessant marginal grafts and captions define what could be called a rhetoric of plagiarism. The insistence on the notion of experience – the rallying-cry of the moderns against the ancients – and the vigour of a language drawn from sailors' slang the better to break down the obtuse and measured rigour of the learned; or the claims for an open air science against the doctrinaire tradition of the schools, were so many manifestations of the new spirit, and at the same time, were the artifices called for by this paradoxical mode of compilation. It is filled with 'negative authorities', used to the precise extent that they are rejected and ridiculed.

At a time when the concept of literary propriety was only beginning to be affirmed – witness the court cases Thevet had brought against him by his 'administrated and coadjutant' associates, or that he in turn initiated – the syphoning-off of earlier or concurrent texts was possible only on condition that one occupied, in the scientific and social fields, a position of strength. Thevet continually railed against the ancients who, not having travelled, had amassed the most 'green' of 'fibs'; or against their modern emulators who, like Belleforest, remained obstinately sedentary and turned their backs on the great discoveries. But that did not suffice to guarantee the authority of the solitary and ubiquitist cosmographer, once he was deprived of the support of the Prince and other powerful mentors. Thevet learned this, to his cost, in the declining years of his career and his life, when there was formed against him a united front of citizens of the Republic of Letters – those adepts in the humanities, whose conversation was erudite and in good taste.

Thevet's failure was not only that of a man unequal to his destiny and ill-prepared for the epistemological revolution in which he was fated to live. He bears witness more generally to the crisis of a genre in transition: at an intermediate stage between medieval *Imagines mundi* and the atlases, encyclopaedias and voyage collections of the classical age. The cosmography renovated by Münster was a momentary and desperately ambitious effort to gather up in a grand synthesis, summarily assembled on the canvas projecting a sphere, the admirable variety of the world. But it would explode after half a century of existence, and give birth to distinct disciplines: the partial knowledges of the topographer, the historian, the botanist, the military engineer, and soon also the statistician.

One has only to open Thevet's *Cosmographie* at random to be struck, as was Flaubert,[10] by the variety of these woodcut illustrations in which monumental inscriptions, in sometimes doubtful Latin, alternate with monsters that renew the meagre substance of medieval bestiaries by adding hasty observations associated with marginal regions or *terrae incognitae*. Thevet's iconoclastic refusal of authorities combined with the uncertainty of distant explorations to unleash a variegated, proliferating imagination. In between obelisks, pyramids and porticoes, the Camphurch, Hulpalim and Haüt escorted bas-reliefs in the ancient style, showing the bruised bodies of Tupinamba athletes dancing, ferociously duelling or engaged in funerary rites. The order of the world, it seemed, was no longer held together by anything but a striated and graduated envelope, artificially offered up to a plethoric mass. The analogical principle of internal cohesion was shattered into a profusion of objects, a surface of marquetry without depth or design. This spectacle, which with Thevet is no longer even subtended by the notion of cosmic harmony, so prevalent in the Renaissance,[11] was unified only by the multiplied gaze of the observer, as it spread around the globe and enveloped it in an immediate grasp. The 'sphere surrounded by the naked eyes that see it', as figured in Thevet's arms, was the precise emblem of the visual possession that henceforward gave form to the universe.

The crisis of cosmography at the end of the Renaissance was manifested, in the final analysis, on three planes. From the religious point of view, the cosmographer who raised himself to the level of the Creator in order to attain the latter's eternal and ubiquitist knowledge was guilty of pride, even blasphemy: he pretended to correct Scripture in the name of his sovereign, unlimited experience. At the level of method, he sinned by incoherence, confusing scales of representation and imagining that autopsy (or seeing for oneself) could guarantee the truth of a synthetic, and necessarily secondary, vision. Finally, from the epistemological point of view cosmography, which supposes a monumental compilation under the uncontrolled authority of a single individual, was soon transcended by more supple and open forms of geographical knowledge. From Giovanni Battista Ramusio to Richard Hakluyt and Théodore de Bry, collections of voyage-accounts whose fortunes were European began, in the late sixteenth century, to supplant a contested genre. Eventually completed by atlases, these collections, in which juridical documents feature side by side with histories of great navigations, prepared the way for the colonial expansion of England, Holland and France.

What, then, was worthy of note in that heterodox and unfinished enterprise, so soon struck with obsolescence? It is, as mentioned, what constituted its very fault: the abundance and variety of material that it assembled into a geometric network as loose as it was possible to get. The work of Thevet and, to a lesser extent, that of his confederate Belleforest incorporate objects that afterwards would escape from science for decades, or even centuries. One would have to wait until Samuel Purchas in the first quarter of the seventeenth century, for example, for the pictograms of the *Codex Mendoza*, from which Thevet drew several of his *Vrais Pourtraits*, to attract the attention of the learned once again.[12] And it was only in the twentieth century that the cosmogony of the Tupinamba Indians or that of the ancient Mexicans, carefully tucked away in Thevet's *Cosmographie universelle*, would at last find adequate readers, in the persons of Alfred Métraux or Claude Lévi-Strauss, of Pierre Clastres or Christian Duverger.[13]

After Thevet the infinite archipelago of the *Grand Insulaire* lay partly submerged; the Malvinas (Falklands) that he was the first to map would remain to be rediscovered, along with the austral Java-la-Grande. The specialization of knowledge that followed the break-up of the cosmographical model brought with it irrevocable 'falls'. There was no longer anywhere to record the unrefined jargon of sailors. Though clear-sighted, Lescarbot, who still accepted Thevet, refused to take seriously the *Voyages avantureux* of Jean Alfonse.[14] Now that manual of pilotage, which mixed very precise nautical instructions with legends and marvels drawn from medieval *Mirrors* – the 'men with tails' in the New World, the lodestone holding Mahomet's body in suspension, or the gold-seeking ants of India[15] – was, as we know, one of the works most often consulted by the cosmographer of four kings.

When what might be called the 'age of cosmography' came to an end, the link was undone between the lowly practical know-how of professional sailors and the refined science of the learned. The possibility of those rudimentary montages of heterogeneous data, those incessant short-circuits between distinct languages, images and sciences by which Renaissance science came to resemble a disconcerting *bricolage*, then vanished. Its art of using up the left-overs of a beleaguered and abused ancient knowledge by mixing in the most incongruous and insolent naïveties was possessed by Thevet in the highest degree. His work bears witness, in a way, to the infinite resources of a 'savage mind' that parasitized the solid compilations of late humanism and forcibly lodged within them, almost by breaking and entering, barbarous material: native words and myths, foreign rites and customs, the raw language of sailors and adventurers, not forgetting, of course, the insistent presence of that irascible and curious commentator himself.

Appendix: Extracts from Guillaume Le Testu's Cosmographie Universelle

---

Unpublished commentaries from the manuscript atlas held at Vincennes (Bibliothèque du Service historique de l'armée de terre: Rés. DLZ 14)

1 Dedicatory epistle of the Cosmographie universelle, to Admiral Gaspard de Coligny, 5 April, 1556 (fols. 1 and 2)

To the High and Mighty Lord Sire Gaspar de Coligny Knight of the Order, Lord of Chastillon, Admiral of France, Colonel of the French infantry, Governor of the Ile de France and Captain of the City of Paris: Guillaume Le Testu, his most humble and obedient servant, wishes peace and eternal felicity.

The great affection I have had, my Lord, to prepare this humble work (which I scarcely judge sufficiently elaborate to be presented to you) has obliged me to submit it, notwithstanding its present rude state; begging you to notice, not the improprieties of its composition, but the good faith in which it is presented to you. Perhaps some will ask, Who is this new cosmographer who, after so many authors both ancient and modern, pretends to discover new things? But I would reply to them, that nature is not so constrained or subjected by the writings of the ancients that she has lost the power to produce new and strange things, beyond those of which they wrote. Although it may be true that the ancients took the greatest pains that were possible for them, they could never have seen all of her effects; or even if they had seen them, they would have been impotent to write them all down; or again, none could have written more than was revealed to him by God. For these reasons, my Lord, I judge that such persons will be satisfied if they consider those who, by their writings, have immortalized their name and that of the people they dedicated their writings to. Accordingly, and trusting that the present volume may be worthy of you and your posterity, I dedicate and present it to you with all my affection; begging you to receive it kindly and, in so doing, to give me the courage that makes a man undertake all that is in his power to accomplish under heaven, for your pleasure. And further, my Lord, I pray that the Creator may grant you ever greater powers, and bring his holy promises to eternal fruition in you. Signed in the French city of Grace [Le Havre] on the fifth day of April 1555, before Easter.

2   West Africa (fol. 21)

This is a part of Africa surrounded by the Ethiopian Sea, situated under the equator in the torrid zone, where some rivers, mountains, and provinces have been described: such as Malcie, Sanaga, Sue and Ethiopia; the kingdom of Mely, Churite; and those of Orguene, Guinée and Benin. The region is fertile and abundant in several kinds of grains, such as milcq [maize], and other grains and fruits. It also abounds in spices such as pepper [*maniguette pouaivres*]. Also, one finds there a great deal of gold; and it has tigers, elephants, lions, ounces,[1] leopards, rhinoceroses and many kinds of beasts and snakes. Among the latter is a serpent that grows to a length of 600 to 700 feet, which serpent eats bullocks and goats; as is reported by Amerigo Vespucci the Florentine in his cosmography of the New World. The inhabitants are black negroes and go naked, with only their shameful parts covered. They regularly make war against each other; and among them are men whose lips are so great that they hang down to their chest, along with others who have only one eye in their forehead, who are marvellous and horrible to see.

3   The Southern Land (fol. 35)

This area is called the austral region, because some say that there is a land to the south, or what is called Auster. However, what I have marked and depicted is only by imagination, and I have not noted or remarked on any of the commodities or incommodities of the place, nor its mountains, rivers, or other things; for there has never yet been any man who has made a certain discovery of it. Therefore I defer speaking of it until we have a more ample report. In the meantime, however, until our knowledge is greater, I have marked and named some promontories or capes in order to align the pieces in which I depict the area.

4   The Southern Land (fol. 39)

This piece is a part of the Southern or Austral Land. It is situated conjecturally in the frigid zone, on account of the fact that many are of the opinion that the land at Magellan's strait and Java-la-Grande are joined together; which is not as yet certainly known. For this reason, I cannot describe anything about the commodities of the latter.

5   Brazil (fol. 45)

This piece demonstrates a part of America, where the regions of brazil [-wood], of cannibals, and the kingdom of Prate [the River Plate] are described. It is situated below the torrid zone, from under the first climate opposite the meridian through Meroe [*antidia meroes*] to the fourth climate opposite Rhodes [*antidia Rodou*]. It is bounded in the north by the ocean of cannibals and the Antilles, and to the east by the great [Atlantic] ocean. All the inhabitants of this land are savages, with no knowledge of God. Those who live upstream near the equator are evil and vicious; they eat human flesh. Those further from the equator, being lowland people, are more tractable. All the said savages, both upstream and down, go naked; their huts and houses are covered with bark and leaves. They regularly make war against each other; that is, those from the mountains fight those from the coasts. The region is fertile in millet and manioc, which is a white root from which they make flour for cooking; for they have no bread. Also, there is a great deal of rape [turnips], of much better quality than in France, along with pineapples, a most delicious fruit, and many other sorts of fruit. Also found in this land are boars, wolves, agoutis, *loups serviers*,[2] armadillos, and many kinds of beasts; together with various sorts of poultry similar to those in France, and parrots with assorted plumages. The merchandise produced in this land includes cotton, brazil-wood, dye-peppers, and large periwinkles from which are made paternosters and women's girdles. The above-mentioned inhabitants are great fishermen, and very skilled with the bow and arrow.

6   Northern Brazil (fol. 47)

This land is the part of America from the northern coast, near the Antilles, to the eastern coast of the cannibals. It is quite convenient and habitable, and situated below the torrid zone from under the first climate opposite Meroe [*antidia meroes*] to the sixth climate opposite Pontus [*antidia Pontou*]. The men of this region go naked, eat human flesh and are very vicious; so much so that one cannot trade with them. They are skilled with the bow and arrow. Their riches include necklaces made of periwinkles and the teeth of the men they have eaten, together with green and white stones which they set into their lips. Also, they have feather ornaments of many colours. They produce much cotton, dye-peppers, and yellow wood. Their food is millet and manioc, from which they make flour to eat; along with rape, pineapples, mahogany nuts, and various sorts of fruits and roots. Also, there are parrots and other birds, along with apes and monkeys, such as are regularly brought from Brazil to this land of France.

7 Patagonia (fol. 50)

This is a part of the land of America near the kingdom of Gingatou. It is bounded to the south by the Straits of Magellan, to the west by the [South?] Sea and the Island of Magellan, and some portion or part of the Austral Land situated below the temperate zone. Its inhabitants are ten to twelve ells [*couldes*] high and speak only by whistling, as Magellan reported. They live on grains such as millet, manioc and several other sorts; as well, there are animals such as armadillos, agoutis and wolves, though not as large as those in France; all of which flesh they eat raw. The region also has parrots and many other sorts of birds with plumages of many colours.

8 Newfoundland as an archipelago (fol. 57)

This is the new-found land, part of the region called 'bacaillaux' [of cod] situated in the temperate zone, under the seventh, so-called *dia boristhenous* climate. From here one can see Canada, Saguenay, Ochelasa and several other places in which the savage men of this place make their home; who go about dressed in various sorts of skins, and believe only in the moon and sun. They are skilled with the bow and arrow, and at throwing harpoons. They get their food by shooting stags, does and many sorts of creatures, both animal and bird; which they dry after they have killed them. They also kill the fish known as sea-wolves, which provide them with oil because they are very fat; and in the said oil they store their provisions, which can last for a year without spoiling.

9 Newfoundland as a continent (fol. 58)

This is a part or portion of the Terra Nova, called the 'Région de bacaillaux' [cod] and known to the navigators of old; who said nothing about the commodities of the place, until Robert Val [Roberval] and Jacques Cartier went there by command of King François, the first of the name. They discovered Canada, and Saguenay, as you saw on the previous page.

Notes

The abbreviation BN indicates the shelf mark or location in the Bibliothèque Nationale, Paris.

Overture: Renaissance and Cosmography

1 Gilbert Chinard, *L'Exotisme américain dans la littérature française au XVIe siècle d'après Rabelais, Ronsard, Montaigne, etc.* (Paris: Hachette, 1911; facsimile edn, Geneva: Slatkine Reprints, 1978).
2 Geoffroy Atkinson, *Les Nouveaux Horizons de la Renaissance française* (Paris: Droz, 1935).
3 See, apart from the work cited of Geoffroy Atkinson, my paper 'Guillaume Postel et l' "obsession" turque', in *Actes du colloque Guillaume Postel (1581–1981)* (Paris: Guy Trédaniel, 1985), pp. 265–98. Reprinted in Frank Lestringart, *Écrire le monde à la Renaissance* (Caen: Paradigme, 1993, pp. 189–224).
4 Stéphane Yérasimos, 'De la collection de voyages à l'histoire universelle: la *Historia universale de' Turchi* de Francesco Sansovino', *Turcica* 20 (1988), pp. 19–41. As the author observes (pp. 21–5), this distinction was made in particular by the Venetian publishers.
5 Girolamo Cardano, *Hieronymi Cardani Mediolanensis Medici De rerum varietate libri XVII* (Basle: apud Henrichum Petri, 1557). See *Opera omnia.* 10 vols (New York: Johnson Reprints and Harcourt Brace Jovanovich, n.d.), Book 12, ch. lx ('Corographicae descriptiones'). I studied this chapter in Yves Giraud (ed.), *Le Paysage à la Renaissance* (Fribourg: Éditions Universitaires de Fribourg, 1988), pp. 9–26.
6 As do the 'Warriors of excise' in Rimbaud's poem 'The Customs Men' (verse 6), 'who slash the blue frontiers with their great axe-blows', in *Rimbaud, Collected Poems*, trans. Oliver Bernard (Harmondsworth: Penguin, 1980), p. 115.
7 See below, ch. 1, nn. 9, 10.
8 As Ptolemy puts it in his *Geography.* See *The Geography of Claudius Ptolemy*, ed. Edward Stevenson (1932), I, 1.
9 For cases illustrating the morphological dynamism of cosmography in the work of the geographers Le Testu and Thevet, see end of ch. 5, below.
10 See Frances A. Yates, *The Art of Memory* (London: Routledge, 1966), ch. 1, 'The Three Latin Sources for the Classical Art of Memory'.
11 Le Testu's method, combining nautical experience with anticipatory 'imagination', is analysed in ch. 5, below.
12 The critique of the notion of 'progress' as a conception foreign to the Renaissance is convincingly developed by Jean Starobinski in *Montaigne in Motion*, trans. Arthur Goldhammer (Chicago: University of Chicago Press, 1985), pp. 279–84.

13 Cf. Atkinson, *Nouveaux horizons*, pp. 289–97: 'Un attardé: André Thevet'. The same epithet and judgement appear again on pp. 428–9.

14 On this *ad hominem* polemic, referred to both here and elsewhere, see my study *André Thevet, cosmographe des derniers Valois* (Geneva: Droz, 1991), chs 7, 8.

15 As outlined by George Huppert, *The Idea of Perfect History: Historical Erudition and Historical Philosophy in Renaissance France* (Urbana: University of Illinois Press, 1970).

16 Atkinson (*Nouveaux horizons*, p. 297) considers that Thevet 'sinned against the truth and opposed his manifestly incredible "observations" to the plausible, and to the facts established by the experience of his time'. The first critic to rectify this caricatural view was Jean Céard, in his thesis *La Nature et les prodiges. L'insolite au XVIe siècle, en France* (Geneva: Droz, 1977), pp. 282–3.

17 Cf. ch. 5, n. 71, below.

18 On this filiation from the *Codex Mendoza* (in the Bodleian Library, Oxford) to Thevet's *Vrais Pourtraits*, see Rüdiger Joppien, 'Étude de quelques portraits ethnologiques dans l'oeuvre d'André Thevet', *Gazette des Beaux-Arts* (April 1978), pp. 125–36.

19 Peter Apian, *Cosmographie* (Paris: Vivant Gaultherot, 1553), fol. 4r. See plate 2 and, on this literal illustration of Ptolemy's definitions regarding geography and chorography, see Svetlana Alpers, *The Art of Describing. Dutch Art in the Seventeenth Century* (London: John Murray, 1983), ch. 4 ('The Mapping Impulse in Dutch Art'), pp. 133–4, 167–8. (Translator's note: Apian's *Cosmographicus liber* (Landshut: by the author, 1524; subsequent edns by Gemma Frisius) is untranslated into English.)

20 See on this point François Secret, 'Notes pour une histoire de l'alchimie en France. 1. André Thevet et l'alchimie', *Australian Journal of French Studies* 9, fasc. 3 (1972), pp. 217–19.

21 Richard Hakluyt, *The Principall Navigations, Voiages and Discoveries of the English nation, made by Sea or over Land, to the most remote and farthest distant Quarters of the earth* . . . (London: George Bishop and Ralph Newberie, 1589), fol. 2r: '[Dedication] To the Right Honorable Sir Francis Walshingham Knight, Principall Secretarie to her Majestie . . .' On the genre of 'cosmographical meditations', see also ch. 2, below.

22 As chaplain to the English ambassador (Sir Henry Stafford), Richard Hakluyt was in Paris from 1583 to 1588.

23 Cf. Thevet, *Cosmographie universelle*, vol. 1, ch. 1, fol. 1r ('What Cosmography is, what it is necessary to observe for an understanding and cognizance of it').

24 This process of construction has been illustrated by Henry Harrisse, in the context of Newfoundland. See his *Découverte et Évolution cartographique de Terre-Neuve et des pays circonvoisins (1497–1501–1769)* (Paris: H. Welter, 1900), p. 278: 'The method of construction was, necessarily, that of all ancient cartographers: that is to say, place-names were added to an ensemble that already existed, often since long before.'

25 Jean Lafond, 'La notion de modèle', in *Le Modèle à la Renaissance* (Paris: Vrin, 1986), p. 12.

26 On this topic, see the convenient synthesis by William G. L. Randles, *De la terre plate au globe terrestre. Une mutation épistémologique rapide (1480–1520)* (Paris: Armand Colin, 1980), pp. 41–64.

27 See Arthur Heulhard, *Villegagnon, roi d'Amérique. Un homme de mer au XVIe siècle (1510–1572)* (Paris: E. Leroux, 1897).

28 See Roger Schlesinger and Arthur P. Stabler, *André Thevet's North America. A Sixteenth-Century View* (Kingston and Montreal: Queen's and McGill University Press, 1986), pp. xxvi–xxxii.
29 For a discussion of this unpublished atlas, see ch. 5.

Chapter 1 The Cosmographical Model

1 For an analogous discussion of the case of Oviedo, which is not unrelated to that of Thevet, see the contribution of Karl Kohut, 'Humanismus und Neue Welt im Werk von Gonzalo Fernandez de Oviedo', in *Humanismus und Neue Welt*, DFG (Deutsche Forschungsgemeinschaft), Mitteilung XV der Kommission für Humanismusforschung (Weinheim: Acta Humaniora, VCH, 1987), pp. 65–88, esp. 69.
2 Gonzales Fernandez de Oviedo y Valdés, *Historia general y natural de las Indias* (Madrid, 1535), II: 1 ('Epistle to the Emperor Charles V'), translated by Jean Poleur as *L'Histoire naturelle et générale des Indes, isles et terre ferme de la grand mer oceane, traduicte de castillan en françois* (Paris: M. de Vascosan, 1555). (Translator's note: the work was not translated into English; both existing versions of Oviedo's work are from his 'Summary' (*Sommario*) of it, a separate work concentrating on natural and social history, rather than the geopolitical aspect. These are *The [natural] hystorie of the weste Indies*, trans. Richard Eden (London, 1555–7; repr. in *Purchas his Pilgrimes*, 1625, and in Edward Arber (ed.), *The First Three English Books on America* (Birmingham: Turnbull and Spears, n.d.), pp. 208–70), and *A Summary of the natural and general history of the West Indies, written by Gonzalo Fernández de Oviedo, alias de Valdés*, trans. and ed. S. Stoudmire (Chapel Hill: University of North Carolina Press, 1959).)
3 Oviedo, *Historia general* (1st edn), Book I (fol. 2v of the Poleur edn). (Translator's note: Oviedo did not discuss his relationship with Pliny in the *Summary* (see n. 2), but did paraphrase the aspects in question here and in nn. 4 and 5. Thus, 'I . . . have had keen desire to learn of these things and I have therefore observed them very carefully. This *Summary* will not reduce the value, as I have said, of the more extensive document which I have already written' (Stoudmire edn, p. 4; see also pp. x–xi.).)
4 Ibid. (Poleur edn, fol. 4r).
5 Ibid. (Poleur edn, fol. 3r).
6 See François Hartog, *The Mirror of Herodotus: The Representation of the Other in the Writing of History*, trans. Janet Lloyd (Berkeley and Los Angeles: University of California Press, 1988), pp. 250–66.
7 Thevet, *Cosmographie universelle*, II, fol. 2r ('Table of the Most Remarkable Things of Europe and of the Fourth Part of the World'):

The Author, being ill, is presented with human flesh for his meal	957.a
The Author is insulted by a Greek priest	816.a
The Author is in danger of his person	491.a
For his second voyage the Author embarks at Saint-Malo	598.b
The Author is in danger of his person	735.b
The Year the Author left Venice	778.a

Or again, in vol. I ('Table of the matters of Africa, and Asia'):

The Author is in danger of his life	300.a
. . .	

The Author is beaten by a doctor	346.b
. . .	
The Author is bludgeoned by a Turk	372.b
. . .	
The Author is imprisoned in the castle at Jerusalem	170.b
The Author is outraged by a Turk	428.b

8   Ibid., I: fol. â3r (end of 'Epistle to the King').

9   On technical aspects of the Treaty of Tordesillas, see the study of Luis de Albuquerque, 'O Tratado de Tordesilhas e as dificuldades técnicas da sua aplicação rigorosa', in the proceedings of the conference *El Tratado de Tordesillas y su Proyección, actas do 1. Colóquio Luso-Espanhol de História de Ultramar* (Valladolid, 1973), vol. I, pp. 119–36.

10 See Pierre Chaunu, *European Expansion in the Later Middle Ages*, trans. Katherine Bertram, vol. 10 of *Europe in the Middle Ages: Selected Studies* (New York: North-Holland, 1979), pp. 179, 132.

11 Pedro de Medina, *L'Art de naviguer de Maistre Pierre de Medine, Espaignol, contenant toutes les reigles, secrets, et enseignemens necessaires à la bonne navigation, traduict de Castillan en François . . . par Nicolas de Nicolai, du Dauphiné, Géographe du tres-chrestien Roy Henri II. de ce nom* (Lyon: G. Rouillé, 1561), 'Proeme', fol. 3r (no published English translation). The colophon indicates that the printing was completed on 2 April 1555.

12 Cf. ibid. and the Book of Solomon 5: 10: '. . . as a ship cleaving the troubled deep, whereof, when it hath gone, no trace can be found, neither the pathway of its keel in the waters', *The Praise of Wisdom, being part 1 of the Book of Wisdom*, rev. and trans., with notes, by E. Blakeney (Oxford: Basil Blackwell, 1937), p. 21.

13 Medina, *Art de naviguer*, II, 1 (p. 25): 'The sea has no condition of colour, properly speaking: for our view does not stop at the surface of the water, but goes deeper: and when one looks across it sideways, it has more or less the colour of the sky: and when winds agitate it, it forms diverse colours.'

14 Thevet, *Cosmographie universelle*, II, Book XXI, ch. 1, fol. 906v (second pagination, wrongly numbered 907).

15 Théodore de Bry, *Americae pars quarta* (Frankfurt am Main: by the author, 1594), plate XV. For two different readings of this engraving of Jean Galle, after a sketch by Stradan, see Bernadette Bucher, *La Sauvage aux seins pendants* (Paris: Hermann, 1977), pp. 214ff, and my study, 'La Flèche du Patagon ou la preuve des lointains', in *Actes du colloque 'Voyager à la Renaissance'* (Paris: G.-P. Maisonneuve et Larose, 1987).

16 Thevet, *Cosmographie universelle*. fol. 975v ('How Mariners Observe the Altitude of the North').

17 Cf. Gustave Flaubert, *The Letters of Gustave Flaubert, 1857–1880.* ed. and trans. Francis Steegmuller (Cambridge, Mass.: Harvard University, Belknap Press, 1982), vol. 2. Cf. also the comment of Pierre-Marc De Biasi, editor of Flaubert's *Carnets* (Paris: Balland, 1988, p. 668), on *Carnet 16, c.*1872: 'A charming image: how mariners observe the altitude of the north – sea, ship, sky all dark and pricked with stars, riddled with unequal white stars.'

18 Gilbert Chinard, *L'Exotisme américain dans la littérature française au XVIe siècle* (Paris, Hachette, 1911), pp. 100–1.

19 Arthur Rimbaud, 'The Drunken Boat' (Le Bateau ivre), verse 32, in *Four Poems by Rimbaud: The Problem of Translation*, trans. Ben Belitt (London: The Sylvan Press, 1948), p. 49.

20 'Géographes et cosmographes de cabinet qui retardent sur les géographes et

les cosmographes de plein vent'. Lucien Febvre, *The Problem of Unbelief in the Sixteenth Century. The Religion of Rabelais*, trans. Beatrice Gottlieb (Cambridge, Mass., and London: Harvard University Press, 1982), ch. 11 ('A Possible Support for Irreligion: The Sciences'): 'These were closet geographers and cosmographers, who lagged behind the open-air geographers and cosmographers'. The question, as we see, is in fact more complex. Progress is not necessarily, and not only on the side of the 'open air'. Cf. also Eric Dardel, *L'Homme et la terre* (Paris: Éditions du CTHS, 1990), pp. 109–15.

21 Bernard Palissy, *Discours admirables, de la nature des eaux et fonteines, tant naturelles qu'artificielles, des metaux, des sels et salines, des pierres, des terres, du feu et des emaux* (Paris: Martin Le Jeune, 1580). For a commentary on these pages, see Lucien Febvre, *The Problem of Unbelief*, pp. 381–2, 385–6. Cf. also François de Dainville, *La Géographie des humanistes* (Paris: Beauchesne, 1940), p. 85.

22 See Marie-Madeleine Fontaine, 'Banalisation de l'alchimie à Lyon au milieu du XVIe siècle, et contre-attaque parisienne', in *Il Rinascimento a Lione. Atti del Congresso Internazionale (Macerata, 6–11 maggio 1985)*, ed. Antonio Possenti and Giulia Mastrangelo (Rome: Edizioni dell'Ateneo, 1988), pp. 263–322.

23 Jean d'Indagine, *Chiromance et Physiognomie par le regard des membres de l'homme, faite par Jean d'Indagine ... le tout mis en françois par Antoine du Moulin, Masconnois, valet de chambre de la Royne de Navarre* (Lyon: Jean de Tournes, 1549) (BN: Rés. V. 2243), pp. 3–10; esp. 5–6, on the parallel with the great navigations, cited by M. M. Fontaine, 'Banalisation de l'alchimie'. The epigraph of the present chapter is the passage in Palissy's *Discours admirables*, in which he sings the praises of pilots expert in the art of navigation.

24 Michel de Montaigne, *The Complete Essays of Michel de Montaigne*. trans. and ed., with intro. and notes, by M. A. Screech (Harmondsworth: Penguin, 1991), III, 13 ('On Experience'). (Translator's note: the French word *expérience* means both 'experience' and 'experiment'. The same double sense is covered by the word *essai*, mentioned here and in the earlier paragraph on Palissy (where it is translated 'essay or assay'); cf. also the title of ch. 5 below. These pointers to the intellectual origins of modern science, lost in modern English, were familiar at the time or soon after in England.)

25 Leonardo Fioravanti, *Lo specchio di scienza universale* (Venice, 1564), cited in the French translation of Gabriel Chappuys, *Miroir universel des arts et sciences*, 2nd edn (Paris: Pierre Cavellat, 1586), I, 39 (p. 167).

26 Thevet, *Cosmographie universelle*, I, Book 1, ch. 2 (fol. 5v). On the contempt Renaissance men often had for their less curious ancient predecessors, see further M. Mollat *Études d'histoire maritime* (Turin: Bottega d'Erasmo, 1977), pp. 667ff, esp. 672–3.

27 Anon., *Le Supplement du Catholicon, ou Nouvelles des Regions de la Lune, ou se voyent depeints les beaux et genereux faits d'armes de feu Jean de Lagny, frere du Charlatan, sur aucunes bourgades de la France, durant les estats de la Ligue. Dedié à la Majesté Espagnole, par un Jesuite, nagueres sorty de Paris* ('Supplement to the Catholicon, or News from the regions of the Moon, in which are depicted the fine and generous military deeds of the late Jean de Lagny, brother of the Charlatan, in various townships of France during the reign of the League. Dedicated to the Spanish Crown by a Jesuit, lately banished from Paris') ([Paris], 1604), ch. 4: 'On a trapdoor that was

opened for us, through which we saw what was happening on the earth'. This opuscule is bound with the *Satyre Menippee de la vertu du Catholicon d'Espagne et de la tenue des Estats de Paris. Derniere Edition* (Paris, 1604) (BN: Rés. Lb³⁵455 D).

28 *Nouvelles des Regions de la Lune*, ch. VII: 'On the Second Quarter of the Moon, from which we were shown the Land of the People from beyond the Water'.

29 This epithet, drawn from the *Epithetes* of Maurice de la Porte, refers to Rabelais. See Christiane Lauvergnat-Gagnière, *Lucien de Samosate et le lucianisme en France au XVIe siècle. Athéisme et polémique* (Geneva: Droz, 1988), ch. VII, pp. 235ff. (Translator's note: see Christopher Robinson, *Lucian and his Influence in Europe* (Greensboro: University of North Carolina Press, 1979).)

30 Agrippa D'Aubigné, *Les Tragiques*, II, vv. 1428–30. See the edition of d'Aubigné's *Oeuvres* by Henri Weber (Paris: Gallimard, 1969), esp. p. 87. (Translator's note: the *Oeuvres* are untranslated into English; see in general Keith Cameron, *Agrippa d'Aubigné* (New York: Irvington, 1977).)

31 D'Aubigné, *Tragiques*, II, v. 1440 (p. 88). For the similar case of the poet John Donne, 'Whose constant allusion to maps and graticules deserves much more attention', see David Woodward, 'The Image of the Spherical Earth', *Perspecta 25. The Yale Architectural Journal* (1989), pp. 3–15, esp. 5.

32 D'Aubigné, *Tragiques*, II, vv. 1431–8. On this reversal of perspective, see Jean Céard, 'Le thème du *monde à l'envers* dans l'oeuvre d'Agrippa d'Aubigné', in *L'Image du monde renversé et ses représentations littéraires et paralittéraires de la fin du XVIe siècle au milieu du XVIIe*, ed. Jean Lafond and Augustin Redondo (Paris: Vrin, 1979), pp. 117–27.

33 D'Aubigné, *Tragiques*, VII, v. 1218 (p. 243).

34 Isabelle Pantin, 'L'Hymne du Ciel', in Madeleine Lazard (ed.), *Autour des 'Hymnes' de Ronsard* (Paris: Champion (Unichamp), 1984), p. 209.

35 Pierre de Ronsard, 'Hymne de la Philosophie', vv. 63–6. See *Oeuvres complètes*, ed. Paul Laumonier (Paris, STFM), vol. VIII ('Les Hymnes de 1555 et de 1556'), p. 89; or *Poems of Pierre de Ronsard*, ed. Nicolas Kilmer (Berkeley: University of California Press, 1979).

36 As noted by Jean Céard in *La Nature et les prodiges. L'insolite au XVIe siècle, en France* (Geneva, Droz, 1977), p. 202. Ronsard was inspired, in passing, by a Homeric image mentioned by Macrobius in his 'Commentary on Scipio's Dream (1, XIV 15–16).

37 Jean Dorat's 'Ode latine' and Guy Le Fèvre de la Boderie's 'Ode pindarique' both feature in the preliminaries to the *Cosmographie universelle* in 1575. They are reproduced in Roger Le Moine (ed.), *L'Amérique et les poètes français de la Renaissance* (Ottawa: University of Ottawa Press ('Les Isles fortunées') 1972).

38 Pierre de Ronsard, *Responce aux injures et calomnies, de je ne sçay quels Predicans, et Ministres de Geneve* (Paris: G. Buon, 1563). See *Oeuvres complètes*, ed. Paul Laumonier, vol. 11, p. 169 (v. 1046). *Poems of Ronsard*, ed. Kilmer, pp. 128–31, has an extract of this poem, not including the line cited. Cf. also Ronsard's *Hymne du Ciel* in *Oeuvres complètes*, vol. 8, p. 142: '. . . for in the round form / Lies the perfection that fully abounds in itself.'

39 Cf. Claude Nicolet, *Space, Geography, and Politics in the Early Roman Empire* (Ann Arbor: University of Michigan Press, 1991), ch. 2, p. 35: 'for the globe

is less the sign of the concrete domination of space easily located on the surface of the earth than of sovereignty the more recognizable for being general and "cosmic" even more than geographic. No empire, no universal monarch could in antiquity reasonably wish to dominate the entire terrestrial sphere.' In Renaissance Europe, the ambition of court geographers would be to make these two conceptions coincide theoretically.

40   The expression is borrowed from Daniel Ménager, *Ronsard, Le roi, le poète et les hommes* (Geneva: Droz, 1979), p. 183.

41   Jean de Léry, *Histoire d'un voyage faict en la terre du Bresil* (Geneva: Antoine Chuppin, 1578, 1580, 1585), preface and passim. See *The History of a Voyage to the Land of Brazil Otherwise Called America*, trans. Janet Whatley (Berkeley: University of California Press, 1990); Lancelot Voisin de La Popelinière, *L'Histoire des histoires. Avec l'idée de l'histoire accomplie* (Paris: M. Orry, 1599), Book 8, pp. 455–9; Richard Hakluyt, *The Principall Navigations*, 1st edn (1589), 'Address to the Reader'; ibid., *The Second Volume* (1599), dedicatory epistle 'To the Right Honorable Sir Robert Cecil, Knight'. On Huppert, see 'Overture', n. 15, above.

42   François de Belleforest, *La Cosmographie universelle de tout le monde ... Auteur en partie Munster, mais beaucoup plus augmentée, ornée et enrichie ...* (Paris: Nicolas Chesneau and Michel Sonnius, 1575), vol. 2, passim; Ludwig Camerarius, letter to Hubert Languet in Vienna, dated Annaberg, 22 July 1575 (BN: ms latin 8583, fol. 49r–v).

43   See Georges Gusdorf, *Les Sciences humaines et la pensée occidentale*, III: *La révolution galiléenne* (Paris: Payot, 1969), vol. 1, p. 128.

44   Dainville, *Géographie des humanistes*, ch. 3, pp. 84ff.

45   These were all blasphemies that François de Belleforest denounced in his rival *Cosmographie* of 1575 (vol. 2, passim). On Thevet's tenacity in defending his perilous statements, see my study mentioned above ('Overture', n. 14).

46   The term was used by Gabriel du Préau, *Histoire de l'estat et succes de l'Eglise dressee en forme de chronique generalle et universelle* (Paris: Jacques Kerver [and G. Chaudière], 1583), vol. 1, 'Epistre' (from fol. â iij). This author stigmatizes 'certain Cosmographers, historians, and poets of our age' for their arrogance with regard to the ancients. By way of a sort of compensation, vol. 2 (published by Kerver's widow, 1583) was preceded by an obsequious 'Advertisement' signed by Thevet.

47   Dainville, *Géographie des humanistes*, p. 87.

48   Montaigne, *Essays*, vol. 1, XXXI ('On the Cannibals'), p. 231.

49   Thevet, *Grand Insulaire*, I, fol. 206r. The rest of the passage violently attacks 'a certain Benzoni and the gentle de Léry, who have glossed and pillaged my history of Antarctic France as much as they were able'.

50   Ibid., fol. 205v.

51   Thevet, *Cosmographie universelle*, I, fol. 6v (my italics). This passage was taken up by Bénigne Saumaize in his commentary on Dionysius the Periegete (Paris, 1597), fol. 47r–v.

52   Thevet, *Cosmographie universelle*, I, fol. 6vff.

53   Ibid., Preface (fol. â 5r).

54   On the case of Empedocles, see the suggestive pages of Gérard Defaux in *Le Curieux, le glorieux et la sagesse du monde dans la première moitié du XVIe siècle* (Lexington, Ky.: French Forum, 1982), pp. 119–28. Defaux recalls, notably, how much the legend of a glorious and sad Empedocles owed to the *Vita* of Diogenes Laertius.

55 I borrow this expression from the title of a collection by Henri Michaux, *Connaissance par les gouffres* (Paris: Gallimard, 1961). (Translator's note: this topographical nuance is somewhat lost in the work's English translation: *Light through Darkness: Explorations among drugs*, trans. Haakon Chevalier (London: Bodley Head, 1964).)

56 Thevet, *Cosmographie universelle*, I, Preface (fol. â 4v).

57 Thevet, *The New Found Worlde, or Antarctike*, trans. Thomas Hacket (Amsterdam: Theatrum Orbis Terrarum; a facsimile of the 1568 London edn by Henrie Bynneman), ch. 19, fols. 29v–31r.

58 Ibid., fol. 31r–v. This is the conclusion of a chapter entitled, 'That not onely all that is under the lyne is inhabited, but also al the worlde is inhabited contrary to the opinion of our elders.'

59 Montaigne, *Essays*, III, 11 ('On the Lame'), pp. 1160–72.

60 Thevet, *The New Found Worlde*, ch. 19 (fol. 31r).

61 See the references in nn. 51, 52 above; also Thevet, *Le Grand Insulaire*, I, fol. 277r.

62 Thevet, *Cosmographie universelle*, II, fol. 910v (S. Lussagnet, *Le Brésil et les Brésiliens par André Thevet* (Paris: Presses Universitaires de France, 1953), p. 25): 'In the first place, let us see where it is that the Indies are placed, and let us see on the globe and round sphere if such an opinion can be sustained (for by the flat sphere you cannot draw any certain judgement) and then we shall see that all the lands of the Indies are Oriental . . .' It was thus a question of showing that America or the 'fourth part of the world' was a separate part of the Indies.

63 Pascal, *Pensées*, trans. A. J. Krailsheimer (Harmondsworth: Penguin, 1966), no. 85. The point was made by Jacqueline Lichtenstein at the conference *Point de vue et distance dans la peinture du XVIIe siècle'* held at Mulhouse on 25 November 1987.

64 Pascal, *Pensées*, no. 84.

65 Ibid.

66 See Thevet, *Le Grand Insulaire*, II, fol. 144r; also my n. 159, 20 to the *Cosmographie de Levant* (Geneva, 1985).

67 On these two *Iles de Thevet*, see ch. 5, below.

68 Thevet. *Cosmographie universelle*, I, fol. 32v.

69 Thevet. *Cosmographie universelle*, II, XXI, 2, fol. 908v (Lussagnet, *Le Brésil*, p. 5). Cf. Pliny, *Natural History*, ch. X. 1: 'Moreover, it is a miracle of nature that this animal indifferently digests anything that is fed to it . . .' (from du Pinet's 1581 translation, p. 376).

70 Thevet, *Cosmographie universelle*, vol. 1, VI: 9 (fol. 176r–v). The passage deserves to be cited in full: 'In this same place the learned Mattioli forgets himself, when he writes that everything thrown into the said sea floats on the surface and does not sink; even if one would cast into it a man bound and garrotted, or something else heavy. I know not who might have spun him such a yarn, given that I saw on five occasions when I was at the said lake the bones and heads of dead horses and camels thrown into it, numbering over a thousand; and, among other things, the living donkey of a Nestorian Christian, with all his equipment, which our porters deliberately threw in to the bottom of it (this, on account of a dispute they had had two hours previously, over a bottle of wine that they had been refused); and as well another who was drunk threw in the boots of his companion, made in Turquesque fashion. All of which things did not fail to sink immediately to the bottom and were lost to view. The first time I was taken there, some

Arabs had killed three of our people and stolen their clothes, and those devils of griffins threw their bodies in; which sank as quickly as would a sounding lead that one throws into the sea.'

71 Thevet, *Cosmographie universelle*, vol. 1, Book I, ch. 1, fol. 1r. The same idea, borrowed from Pliny, appears in Belleforest's *Histoire universelle* (Paris, 1570), 'Preface to the Reader'.

72 It was almost in these terms that the German editor Simon Grynaeus expressed himself, in his preface to the famous *Novus Orbis* (Basle, 1532). See Jean Céard, *La Nature et les prodiges* (Geneva: Droz, 1977), p. 273; and, on the same collection of voyage accounts considered in terms of its political function, the contribution of Michel Korinman, 'Simon Grynaeus et le *Novus Orbis:* les pouvoirs d'une collection', in J. Céard and J.-Cl. Margolin (eds), *Voyager à la Renaissance* (Paris: G. P. Maisonneuve et Larose), 1987, pp. 419–31.

73 Such was the title, as is well known, of one of Jean Bodin's last works: *Universae naturae theatrum. In quo rerum omnium effectrices causae, & fines contemplantur*... (Lyon: J. Roussin, 1596. It was published in French translation by François de Fougerolles (Lyon: Jean Pillehotte, 1597).

74 Thevet, *Cosmographie de Levant*, 'Epitre' (p. 5, lines 10ff). Cf. his *Cosmographie universelle*, I, Preface (fol. â 5r). The passage was repeated verbatim twenty years later.

75 Thevet, *Cosmographie de Levant* (as n. 74). Cf. his *Cosmographie universelle*, I, fol. â 5r: 'You will read at times about histories, at times about natural questions that are no less true than delectable, within the limits of my humble mind' (or, as the *Cosmographie de Levant* puts it, '... no less delectable than true'). (Translator's note: the French word *histoire* means both story and history. Writers often played on this double meaning to emphasize the narrative or even fictional aspects of history, in a way that runs counter to the resources of modern English and, in general, to modern historiographical notions.)

76 As is the case with Hippocrates at Cos, Aristotle in Euboea, Cicero at Zante (Zakynthos), and so on.

77 This 'play of nature' haunted the thought of Cardano, who was almost Thevet's contemporary. See Céard, *La Nature*, p. 236.

78 See Michel Foucault, *The Order of Things: An Archeology of the Human Sciences*, trans. Alan Sheridan (New York: Pantheon, 1972), Preface.

79 Thevet, *Cosmographie de Levant*, 'Epitre', p. 5.

80 Thevet, *Cosmographie de Levant*, p. 158 (engraving).

81 As recalled, notably, by François Rigolot, *Poétique et Onomastique. L'exemple de la Renaissance* (Geneva: Droz, 1978).

82 See below, ch. 5.

Chapter 2   Ancient Lessons: A Bookish Orient

1 For a typology of voyage accounts in the Renaissance, see the *Actes* of the conference *Voyager à la Renaissance* held at Tours in 1983 (Paris: Maisonneuve et Larose, 1987), in particular, the introduction by Jean-Claude Margolin and concluding remarks by Jean Céard.

2 Lodovicus Coelius Rhodiginus, *Lectionum Antiquarum Libri triginta, recogniti ab Auctore, atque ita locupletati, ut tertia plus parte auctiores sint*

redditi: qui ob omnifariam abstrusarum et reconditiorum tam rerum quam vocum explicationem (quas vix unius hominis aetas libris perpetuo insudans observaret) merito Cornucopiae, seu Thesaurus utriusque linguae appellabuntur, quod in quocumque studiorum genere, non minor ipsorum, quam ingentis bibliothecae, aut complurium commentariorum, possit esse usus (Basle: Henry Frobenius and N. Episcopus, 1542), in-fol., 1183 pp. Also consulted here was a textually identical Geneva edition of 1620, 'Excudebat [printed by] Philippus Albertus'.

Among the hundred-odd references to Coelius Rhodiginus in the 'annotations and scholia' to my edition of Thevet's *Cosmographie de Levant* (Geneva, 1985), two are erroneous: p. 275, n. 71, 17–21 should read: 'Coelius Rhodiginus, IV, 12, p. 132' (not IX, 12); and p. 326, n. 193, 20 should read on the last line: 'The adage is also in Coelius Rhodiginus, XXV, 25' (not XVIII, 34).

To conclude this note on my research in progress, I would mention certain other borrowings by Thevet from the author of the 'Ancient Lessons'. In ch. 9 (p. 36, lines 17–21), the passage on the herb Asplenon, which relieves problems of the spleen (see n. 66, below), comes from *lectio* IV, 18, as was noted by Gregor Horst. In ch. 11, p. 44, lines 14–16, the remark on the low reputation of the people of Chios (see n. 69) is a probable allusion to *lectio* XXVI, 33, also according to Gregor Horst. Again, ch. 22 ('On Camels') can be compared to *lectio* VII, 18 (Horst, p. 68).

3 See the reference to this translation in n. 54, below.
4 See Gaetano Oliva, *Celio Rodigino, saggio biografico dell'età del Rinascimento* (Rovigo: Comune di Rovigo, 1868).
5 Niccolò Perotti (Sipontinus), *Cornucopiae, sive Commentariorum linguae latinae libri*. Venice ('in aedibus Aldi et Andreae Soceri'), 1513 (1st edn, 1489); Ambrogio Calepino, *F. Ambrosii Calepini . . . Dictionarium, ex optimis quibusdam authoribus studiose collectum, et recentius auctum et recognitum* (Paris: J. Badius Ascensius, 1514).
6 See n. 2, above: this was a publicizing comment included in the title of Frobenius's edition of the 'Ancient Lessons'.
7 See the *Fifth Book* of Rabelais (manuscript final chapter), ed. G. Demerson (Paris: Éditions du Seuil, L'Intégrale, 1973), p. 916, n. 14. Rabelais used Rhodiginus's *lectio* XXII, 4 for ch. 47 of his *Fifth Book*. As for Montaigne, cf. Villey's edition of the *Essais*, 3rd edn (Paris: Presses Universitaires de France, 1978), p. lxvi ('Appendice au catalogue des livres de Montaigne').
8 Pierre Boaistuau, *Brief Discours de l'excellence et dignité de l'homme* (Paris: Vincent Sertenas, 1558). See the critical edition by M. Simonin (Geneva: Droz, 1982), esp. nn. 35, 69, 107, 114 and 127.
9 Estienne Tabourot, *Les Bigarrures du Seigneur des Accords (premier livre)* (Paris: Jean Richer, 1588), ch. 15 ('Des acrostiches'), fol. 152 C. See the notes by F. Goyet to his facsimile edition (Geneva: Droz, 1986), vol. 2, p. 124. Cf. also *Lectiones antiquae*, XXV, 9 ('Acroteria, Acrostichis'). Tabourot's reference to *Lectio* XIII, 17 is erroneous. Finally, another example of the fortunes of Coelius Rhodiginus in France: we know that Ambroise Paré, in his book *Des monstres et prodiges* ('On monsters and prodigies', 1573), cites on numerous occasions the 'Ancient Lessons', which offer a wealth of teratological cases. See the edition by Jean Céard (Geneva: Droz, 1971), trans. Janis L. Pallister as *On Monsters and Marvels* (Chicago: University of Chicago Press, 1983).

10 I base myself here on Bruno Rech, 'Bartolomé de Las Casas und die Antike', *in Humanismus und Neue Welt*, ed. Wolfgang Reinhard for Deutsche Forschungsgemeinschaft (Weinheim: Acta Humaniora (VCH), 1987), pp. 168–9.

11 Thevet, *Cosmographie de Levant*, ch. 47, p. 171. See François Secret, *Les Kabbalistes chrétiens de la Renaissance* (Paris: Dunod, 1964), p. 317.

12 Thevet, *Cosmographie de Levant*, ch. 27, pp. 87–90. The *lectiones* cited are, in order of appearance, *Lectiones antiquae* XVIII, 5; XVIII, 2; XVIII, 1; XXII, 23; XVIII, 4; XIII, 6.

13 Thevet, *Cosmographie de Levant*, ch. 11, p. 44, lines 28–30. Cf. Montaigne, *Essays*, I, 23b.

14 Thevet, *Cosmographie de Levant*, ch. 1, pp. 16–17. On the probable construction of this chapter, see the note in my edition of the work, p. 254.

15 On the Renaissance genre of commentary see Jean Céard's study, 'Les Transformations du genre du commentaire', in *L'Automne de la Renaissance (1580–1630)* (Paris: Vrin, 1981), pp. 101–15.

16 These are all topics discussed, after Coelius Rhodiginus, in the *Cosmographie de Levant*.

17 Thevet, *Cosmographie de Levant*, ch. XXV, p. 123, lines 3–4.

18 Cf. Cicero, *Pro Sex. Roscio Amerino* XXV.70; XXVI.71. See Cicero, *Orationes*, ed. A. C. Clark, 6 vols (Oxford: Oxford University Press, 1905), vol. 1; and *The Poems of Catullus*, trans. F. Raphael and K. McLeish (London: Cape, 1978).

19 Thevet, *Cosmographie de Levant*, 'Epitre', p. 5, lines 12–15. Cf. ibid., ch. 7, p. 29, lines 9–12, where a topical comparison is made with honey-flies gathering their booty here and there.

20 *Cosmographie de Levant*, ch. XLVII, p. 171, line 11.

21 Ibid., ch. XL, p. 148, lines 20–2.

22 Ibid., ch. XV, p. 54, lines 10–12. Taken from *Lectiones antiquae*, XXVI, 29–31.

23 Thevet, *Cosmographie de Levant*, chs XVIII, XXI and XXII respectively.

24 Ibid., ch. XXXIX, p. 145, line 23. For the passage in Poliziano's *Miscellanies* on the giraffe, see Angelo Ambrogini (known as Angelo, or Agnolo, Poliziano), *Angeli Politiani operum tomus primus* (Lyon: Sebastian Grypheus, 1539), pp. 515–17.

25 Thevet, *Cosmographie de Levant*, ch. XXXIX, p. 145, lines 24–8.

26 Pliny, *Natural History*, 11 vols (Cambridge: Mass.: Harvard University Press, Loeb Classical Library, n.d.), vol. 8, XVIII.69.

27 Thevet, *Cosmographie de Levant*, Preface, p. 14, lines 1–4.

28 Barthélemy Aneau, *Decades de la description, forme, et vertu naturelle des animaulx, tant raisonnables, que brutz* (Lyon: Balthazar Arnoullet, 1549) (BN: Rés. Ye 3468-1), 'Praeface', fol. Aijr. (Translator's note: this work is untranslated: cf. the similar *Alector. The Cock. Containing the first part of the most excellent amd mytheologicall historie of the valorous Squire Alector; sonne to the renowned Prince Macrobius France-Gal and to the peerelesse Princesse Piscaraxe, Queene of high Tartary* (London: Thos. Orwin, 1590).)

29 See the valuable contribution of Marie-Madeleine Fontaine, 'Alector, de Barthélemy Aneau, ou les aventures du roman après Rabelais', in *Mélanges sur la littérature de la Renaissance à la mémoire de V.-L. Saulnier* (Geneva: Droz, 1984), pp. 563–4.

30 Aneau, *Décades*, fol. A 1v.

31 On the theory of the emblem and its origins, see the collective work by

Claudie Balavoine, Yves Giraud et al., *L'Emblème à la Renaissance* (Paris: SEDES-CDU, 1982), also *Emblematica: An Interdisciplinary Journal for Emblem Studies* (New York: AMS Press, 1987–), passim.

32 Desiderius Erasmus, *Adagiorum chiliades quattuor* (Basle: Frobenius and Episcopus, 1540), I, 5, 93. See *Adages (One to Five Hundred)*, trans. Margaret Phillips and R. A. B. Mynors, vol. 31 of *The Collected Works of Erasmus* (Toronto: University of Toronto Press, 1982); or *Proverbs or Adages*, trans. Richard Taverner (London: Will. How, 1569; reprinted Springfield, Va.: Scholasticus, 1977). Cf. also Rhodiginus, *Lectiones antiquae*, VII, 22.

33 Thevet, *Cosmographie de Levant*, ch. XII, p. 49, line 21 to p. 50, line 8.

34 Ibid., ch. XXXIV (p. 119, lines 8–9). Cf. Rhodiginus, *Lectiones antiquae*, XXIII, 14.

35 Thevet, *Cosmographie de Levant*, ibid., lines 5–6; based on *Lectiones antiquae*, XXIII, 16.

36 Erasmus, *Adages*, IV, 4, 39. Cf. Aulus Gellius, *Noctes Atticae* XVI.9.

37 Thevet, *Cosmographie de Levant*, ch. XIV, p. 52, lines 14–20.

38 Gerardus Mercator and Jodocus Hondius, *Gerardi Mercatoris Atlas, sive Cosmographicae Meditationes de fabrica mundi et fabricati figura. Jam tandem ad finem perductus quam plurimis aeneis tabulis Hispaniae, Africae, Asiae et Americae auctus ac illustratus a Judoco Hondio* (Amsterdam: C. Nicolaius and J. Hondius, 1607), trans. Henry Hexham, as *Atlas: or, A geographicke description*... (Amsterdam: Henricus Hondius, 1630).

39 Sebastian Münster, *Cosmographiae universalis Libri VI*... (Basle: Henricus Petrus, 1550; *La Cosmographie universelle de tout le monde*, Paris: Michel Sonnius and Nicolas Chesneau, 1575), vol. 1, Book I, ch. 2 ('On the Division of the Sea, and the Source of Rivers'), col. 9: 'for you to understand things clearly, you must have in your hands a universal table of the description of the world, which, being before your eyes, will show you all of this, with the various dispositions of the land and the sea. You will see there how the Ocean extends into various gulfs, makes such approaches and so hounds the seas that are enclosed in the lands, that the gulf of Arabia is not further from the Egyptian sea than 115 thousand strides ... [and] how much it occupies the land, and the measure of so many rivers and swamps, to which will be added lakes and ponds, and all those places that have no inhabitants. I will say nothing of the fact that the earth in places rises up to the heavens and has such high summits to see; as well as so many forests, steep valleys, deserts and other places that are not inhabited for a thousand reasons; and yet it is here that lies the subject of our glory. We have here our honours, exercise our empires and seek riches. Here men are troubled and disturbed. We constantly revert to civil wars here, and in killing each other cause the earth to become more vague and less inhabited, for the sake of an empire that lasts only a single hour. O what folly!'

40 Joachim Vadianus, *Epitome trium terrae partium, Asiae, Africae et Europae compendiariam locorum descriptionem continens, praecipue autem quorum in Actis Lucas, passim autem Evangelistae et Apostoli meminere. Ab ipso Authore diligenter recognita, et multis in locis aucta* (Zurich: Froschauer, 1548). The first edition was published in 1534, but it was the 1548 text that Thevet used for his *Cosmographie de Levant*.

41 Thevet, *Cosmographie de Levant*, ch. XLIV, p. 164, line 6; after Vadianus, *Epitome*, fol. 110.

42 Thevet, *Cosmographie de Levant*, ch. L, p. 181, line 30.

43 Ibid., ch. LII, p. 188, lines 22–31.

44 Ibid., ch. XLIII, p. 161, lines 12–15. According to Vadianus (*Epitome*, fol. 77r–v), 'Parum quidem laudata haec veteribus, ob squalorem et sterilitatem, nobis rerum vere divinarum memoria ita frequens et celebris existit, ut multis nominibus odoriferae illic ac diviti et Beatae quam gentes extollunt, praeferri et anteponi debeat. Nam ut summas tantum rerum attingamus, Israelem sicco pede Rubrum mare transgressum prima omnium hospitali benignitate excepit, nec minus quadraginta annis tenuit . . . Vere autem o vere fertilem Arabiam istam credentibus, Arabiam inquam Petream, quae petram, id est, Christum ita nobis adumbravit . . .' This eulogy of Arabia Petrea (Stony Arabia) by Vadianus is transported by Thevet to Arabia Deserta, as a result of a geographical confusion. But the general sense remains unchanged.

45 Thevet, *Cosmographie de Levant*, ch. XV, p. 54, lines 17–18.

46 Ibid., ch. XXI, p. 71, lines 17–21. Cf. *Lectiones antiquae*, IV, 12: 'Genua elephanti non flectunt.' On the demonstration given by Pierre Gilles of Albi in his *Descriptio nova Elephanti* (Hamburg: The Heirs of Philipp de Ohr, 1614), see my edition of the *Cosmographie de Levant*, nn. 71, 22.

47 Thevet, *Cosmographie de Levant*, ch. XXI, p. 72, lines 11–19; after *Lectiones antiquae*, XIII, 18: 'Dentes'ne an cornua dici in elephantis debeant.' The comparison would be made by Gregor Horst (p. 65). See n. 54, below, referring to this work.

48 Thevet, *Cosmographie de Levant*, ch. XXXI, pp. 106–9.

49 Raffaello Maffei (Volaterranus), *Commentariorum urbanorum octo et triginta libri* (Basle: Frobenius, 1530), X, fol. 117v: 'It was to those of Colossae, the town by the River Lycus mentioned by Herodotus [and] by our St Jerome, therefore, that the epistle of Paul was addressed, and not to the people of Rhodes (as the vulgar think).

50 Thevet, *Cosmographie de Levant*, chap. XXIV, p. 81, lines 30–82, 3. This was a classic question, discussed notably by Pliny (IX.6.17–19), whom Thevet follows, and who refuted Aristotle's *Historia animalium* II.15.506a.

51 Thevet, *Cosmographie de Levant*, ch. XV, p. 54, lines 23–7. For another example of balanced judgement, cf. ibid., ch. XI, p. 44, lines 12–17, on the honesty of the inhabitants of Chios: 'It is impossible to please all, for what one approves of, the other condemns: so great is the inconstancy of human beings.' Coelius Rhodiginus voiced serious doubts about the morality of these islanders: see n. 69 of the present chapter.

52 See on this subject Bernard Guenée, *Histoire et culture historique dans l'Occident médiéval* (Paris: Aubier, 1980), pp. 168–9, 303–4.

53 Paris: Chesneau and Sonnius, 1575, vol. I. The title already contains this condescending restriction: 'Authored in part by Münster, but greatly augmented, ornamented and enriched by François de Belle-forest, Comingeois . . .'

54 *Cosmographia Orientis, das ist Beschreibung desz gantzen Morgenlandes, vor diesem Frantzösisch beschreiben, durch* Andream Thevetum, *jetzo aber in Teutsche Sprache versetzt und mit nützlichen* marginalibus *vermehret durch Gregor Horst, der Artzney Doctorn und Professorn, auch Fürstl. Hessischen Leib medicum* (Giessen: Caspar Chemlin, 1617), 4 fols. (BNU, Strasbourg: D 165065; Van Pelt Library, University of Pennsylvania: 915/T. 358, incl. 12 additional plates copied from illustrations in the 1556 edn; Houghton Library, Harvard University: US 2530.2.6*, incl. 2 additional plates numbered IX and XII.).
	This doctor, Gregor Horst the Elder, was also the author of a *Herbarium Horstianum, seu de selectis plantis et radicibus libri duo* (Marburg: C. Chemlin,

1630); of *De Natura humana libri duo, quorum prior de corporis structura, posterior, de anima tractat* (Frankfurt am Main: E. Kempfer, 1612); and three volumes of *Opera medica* published by his sons Gregor Horst the Younger and Johann Daniel Horst (Gouda and Amsterdam, 1661). In a note to ch. 6 of the *Cosmographia Orientis*, 'Von der Verwandlung' ('On Transformations'), the translator obligingly refers the reader to his own books: 'Davon besiehe opus nostrum *de natura humana lib*. 2, exer. 10 quaest. 9.' Here the doctor falls in with Thevet's view on the thorny question of lycanthropy: 'Die jenige welche vermeinen sie sind in Wölffe verändert, haben allein ein [*sic*] falsche Einbildung' (Those who claim to have been transformed into wolves simply have a corrupted imagination), ibid., p. 17.

55 Joachim Strüppe (Struppius), *Consensus celebriorum medicorum, historicorum et philosophorum, super secretiss. ac preciosiss. quibusdam Medicinis fere exoticis, primumque super MUMIA eique cognatis, maxime in Iudaea, Aegypto, Arabia, etc., olim usitatissimis, Tractatus primi ΠΕΡΙΟΧΗ [Perioché]. Ubi Mumiae genuinae, ad Pyramides ex concameratis Caemiteriis Aegyptiacis erutae VERUM EXEMPLAR, inter multa millia, ut carum valde, sic venustissimum, simul et vetustissimum, annis circiter 2000. reconditum, cum admiratione coram aspiciendum exhibetur... Per Ioach. Strüppe de Gelhausen etc. D.* (Frankfurt am Main: N. Bassaeus, 1574) (BN: 4°Te[139]11), 8 fols. in-quarto.

The woodcut entitled 'Expressa verae Mumiae Ægyptiacae effigies', appearing at the end of this booklet (fol. B 4v) was reproduced by Gregor Horst on p. 147 of the *Cosmographia Orientis* to illustrate the chapter (XLII) treating 'Of the Tombs of the Egyptians, Mummies, and Balm'.

This question of the conservation of cadavers could not have failed to interest Gregor Horst, who in 1608 published a Σκέψις *[Skepsis] de naturali conservatione et cruentatione cadaverum, ubi ex casu quodam admirando et singulari duo problemata... deducuntur* (Wittenberg: M. G. Müller).

56 'In all things the Author demonstrates as surely as a pious Christian that in all our needs we should seek no other helper and mediator than God alone, the Almighty', *Cosmographia Orientis*, ch. XI, p. 39.

57 *Cosmographia Orientis*, ch. LII ('On Antioch'), p. 185 (headliner): 'Dieser Autor ist mit der Jesuiten Namen nicht zufrieden' (This Author is not satisfied with the name of the Jesuits). Cf. Thevet's *Cosmographie de Levant*, p. 189, line 8, and my note on the same.

58 'That the Greeks in many ways come closer to the Evangelical Truth than those attached to the Pope in Rome is obvious to all, even though the Author regards it as an error', *Cosmographia Orientis*, ch. XXIX, p. 91, headliner.

59 Ibid., cap. XL, p. 142: 'Dieser Text wird von D. Luthero vertizt, die wechter nach der Version der *Septuag. Interpretum*, da stehet φύλακες sonsten hat der hebreische text *gammadim*, das gebe *Pygmaei*, welche aber ungleich verstanden werden, wie bey dem *Schindlero* und *Buxdorfio* in den Lexicis zu sehen' (This Text is treated by Dr Luther, on the subject of the guards according to the Version of the *Septuag. Interpretum*. There it says φύλακες [*phylakes*], whereas the Hebrew text has *gammadim*, meaning 'Pygmies'; which are not the same thing, as can be confirmed by consulting the Lexicons of Schindler and Buxdorf). The Lutheran Gregor Horst refers here to Buxdorf, *Lexicon hebraicum et chaldaicum*, 2nd edn (Basle, 1615), and to Valentin Schilder, *Lexicon pentaglotton, hebraicum, chaldaicum, syriacum, talmudico-rabbinicum et arabicum* (Frankfurt, 1612).

60 *Cosmographia Orientis*, fol. iiir: 'Dieweil aber unter allen andern, welche die

Reisen zum heiligen Grabe beschrieben, nicht leicht einer gefunden wird, in welchem so kürzlich und mit wenigen worten so viel denkwürdige sachen verfasset' (But among the many works that describe travels to the Holy Sepulchre, it is not easy to find one in which so briefly and with few words so many noteworthy things are assembled).

61 Ibid.: 'Ja es bleibt bey den Historien nicht, sondern kommen zu offterm feine lustige *quaestiones* und Fragen darzu, damit der Leser so viel mehr nutzen darvon habenmöge, zugeschweigen der *Antiquiteten*, so hierben, der Warheit nach, künstlich abgebildet senn' (Indeed, it does not stop at the stories, but often brings in fine amusing *quaestiones* and problems from which the reader might derive so much more benefit; so that in this way the *Antiquities* are portrayed artfully, after the truth).

62 *Cosmographia Orientis*, ch. XI ('On the Island of Chios'), p. 34 (marginal headliner).

63 In 1566. See ibid., pp. 34, 36 (headliners).

64 See for example the *Cosmographie Orientis*, ch. XIV, p. 45: 'Erasmus Roterodamus setzt in seinen *Adagiis*, dass man sagt, *Athos obumbrat latera Lemniae bovis*, wenn einer einem beschwerlich ist, um eines anderen ruhm gedencket zu hindern, gleich wie der Berg Athos in Thracia mit einem Schatten die weisse Seule einer auffgerichten Kuhe in Lemno sehr weit darvon verfinstert' (Erasmus of Rotterdam states in his *Adages* that people say *Athos obumbrat latera Lemniae bovis* when one man is burdensome to another, in such a way as to deprive him of glory; just as Mount Athos in Thracia could darken with its shadows a white cow standing far away on the island of Lemnos). Thevet cites this 'common proverb' (*Cosmographie de Levant*, XIV, 52, 2) without mentioning his debt to Erasmus's *Adages* (III, 2, 90).

65 *Cosmographia Orientis*, ch. IX ('On Candia'), p. 27, note *e*.

66 Ibid. Cf. *Lectiones antiquae*, IV, 18 ('Asplenon medicamentum quid').

67 *Cosmographia Orientis*, ch. XXI ('On Elephants'). Reference is made here (p. 65) to 'Caelio Rhodigino lib. 13 lect. antiq. c. 18', the likely source for the passage. See my edition of the *Cosmographie de Levant*, n. 72, pp. 11–19.

68 *Cosmographia Orientis*, ch. X, p. 31; cf. *Lectiones antiquae*, XXX, 27 ('Terrae motuum species').

69 *Cosmographia Orientis*, ch. XI, p. 35 (headliner). Cf. *Lectiones antiquae*, XXVI, 33 ('De Chiorum turpitudine, obiter et Cappadocum'). This passage, in which the legendary morality of the inhabitants of Chios is denied, seems to account for the allusion Thevet makes in the *Cosmographie de Levant* (XI, 44, 14–16): 'Now I do not want to forget the honesty of the inhabitants of the Island of Chios. It is true that some authors have noted them to be lascivious and dishonest; but it is impossible to please everyone . . . ' Once again, Gregor Horst shows that he probably saw through this; but he gives an incorrect reference to the 'Ancient Lessons' (XXVI, 23 instead of XXVI, 33).

70 *Cosmographia Orientis*, ch. XVIII, p. 61 (headliner): 'Besiehe Cael. Rhodiginum und Aelianum' (see Coelius Rhodiginus and Aelian). At the end of the same chapter on lions, Gregor Horst mentions several adages that Thevet had overlooked, and that could be found in the *Chiliades* of Erasmus: 'Allhie sind noch zu mercken feine Sprichwörter von den Löwen genommen, als *Leonem ex unguibus cognoscimus*, das man sagt, wann man etwas gar gewiss macht . . . Item *Leo risit* . . . Item *Leonem subula non excepit*' (In

various places can be noted fine adages drawn from the habits of lions, such as *Leonem ex unguibus cognoscimus*, which one says when one does something very certain . . .).

71 This presence of the commentator becomes noticeably rarer as the work progresses: for whereas the margins of the first chapters are filled with almost continuous annotations, these become spaced out from ch. XXIX on, and by the end have become very rare.

Chapter 3 Mythologics: The Invention of Brazil

1 Thevet, *Vrais Pourtraits*, II, VIII (chs 149, 145 respectively).
2 On this work (BN: Ms fr. 15454), see Lussagnet, *Le Brésil*, pp. 237–310.
3 Thevet, *Le Grand Insulaire* (BN: Ms fr. 15452), maps 92–5, 98–101. No. 92 (fol. 225v) is 'Isles de Maquehay' (Macaé); no. 93 (fol. 228v) is 'Isle des Margajas' (Ilha do Governador, in the bay of Rio); no. 94 (fol. 232v) is 'L'Isle Henrii (Ilha de Villegaignon, in the bay of Rio)'; no. 95 (fol. 245r) is 'Gouffre de la riviere de Ganabara ou Janaire' (bay of Rio de Janeiro); no. 98 (fol. 259 bis) is 'L'Isle de Thevet'; no. 99 (fol. 262v, mq.) is 'Isles honestes'; no. 100 (fol. 264v, mq.) is 'Isles de la Baye des Roys' (Angra dos Reis); and no. 101 (fol. 265v, mq.) is 'Isle de S. Sebastien' (I. de São Sebastião).
4 Thevet, *Cosmographie universelle*, vol. I, fol. 116r and vol. II, fol. 941r.
5 Ibid., I, II, ch. 16 (fol. 64v). The passage was copied by Louis Guyon de la Nauche in *Les Diverses Leçons* (Lyon: Claude Morillon, 1604, 1610), ch. 28, p. 748.
6 Thevet, *Cosmographie universelle*, I, V, ch. 1 (fol. 121r).
7 Thevet, *New Found Worlde*, ch. 6, fol. 106v.
8 Thevet, *Cosmographie universelle*, II, XXI, ch. 3 (fol. 911v; 2nd pagination). Cf. Lussagnet, *Le Brésil*, p. 29.
9 On this proverb, see Guy Turbet-Delof, *L'Afrique barbaresque dans la littérature française aux XVIᵉ et XVIIᵉ siècles* (Geneva: Droz, 1973), p. 42. On the antagonism of Turk and Moor, see further my communication 'Guillaume Postel et l'obsession turque' in the *Actes* of the conference *Guillaume Postel (1581–1981)* (Paris: Guy Trédaniel, 1985), p. 265ff, esp. 270–1 and 281–2 ('Bon Turc et More cruel').
10 Thevet, *Cosmographie universelle*, I, VI, 1 (fol. 151v).
11 Ibid., ch. 5 (fol. 163r–v).
12 On Melusin (of Lusignan, near Poitiers) see *New Found Worlde*, ch. 81 (fol. 132r); on 'the English prophet Merlin', *Cosmographie universelle*, II, XXI, ch. 6 (fol. 919r) (Lussagnet, *Le Brésil*, p. 67). On 'la conqueste du sainct Graal en la grande Bretaigne', see *Cosmographie universelle*, fol. 920r (Lussagnet, *Le Brésil*, p. 71).
13 Cf. the title of ch. IV of Book XXI of the *Cosmographie universelle*: 'On Cape Frio, and the Slight Beliefs of the Savages of that Land'.
14 Thevet, *Cosmographie universelle*, II, XIX, 10 (fol. 851r): 'For some adored Fire, others the Forests, others Snakes, others the Sun, and a Hammer of monstrous height and thickness. And when they were asked why they so honoured this Hammer, they replied that once upon a time the Sun had gone for a long time without showing them its clarity; but when at last it appeared, the King of the land seized it and imprisoned it in a very strong tower; but the signs of the Zodiac, coming to the prisoner's aid, broke down

this tower with such a Hammer and set the Sun free; and for this benefit they worshipped the Hammer that had thus obligated men by giving them back the clarity of the Sun.'

15 Ibid., II, XVI, 11 (fol. 674r): 'Moreover, when the sea is frozen there, and the ice breaks up, it makes a sound like a human voice; which is why the simple and rude people of that land believe that these are the souls of the dead, who lie there in torment and spend the time of their penitence; such is the belief of that nation in the notion of Purgatory.'

   After Olaus Magnus, *Historia de gentibus septentrionalibus, earumque diversis statibus, conditionibus, moribus, ritibus, superstitionibus, disciplinis, exercitiis, regimine, victu, bellis, structuris . . . et rebus mirabilibus* ([Rome], 1555), Book II, ch. 3 ('De apparentibus umbris submersorum'): 'Ibique [= in Islandia] locus esse creditur poenae, expiationisque sordidarum animarum. Illic nempe spiritus, seu umbrae, comperiuntur se exhibentes manifestos humanis ministeriis submersorum, sive alio violento casu enectorum.' Cf. the epitome in *History of the Northern Lands* by Christopher Plantin (Antwerp, 1561): 'There are in this island some very admirable, and seemingly miraculous, things. Among others, there is a rock that burns and flames continuously, like Mount Aetna (or Mongibel), and which the inhabitants of this land believe is a gulf of hell where the souls of the dead do their penitence. And they say they regularly see souls there, and that spirits are manifestly seen . . .' (translated from the French edition: II, 3, fol. 20r–v).

   The passage in question from Thevet's *Cosmographie universelle* can be related to the Rabelaisian myth of frozen words (*Quart Livre*, chs 55–6), to which the cosmographer would later reply in the *Grand Insulaire* (I, fol. 147r–v). See Roger Schlesinger and A. P. Stabler, *André Thevet's North America: A Sixteenth-Century View* (Montreal: McGill and Queen's University Press, 1986), pp. 235–6.

16 Thevet, *Cosmographie universelle*, II, XXI, 1 (fol. 905v–906r) (second pagination).

17 Ibid., II, XVI, ch. 11 (fol. 674r).

18 See Elfriede Regina Knauer, *Die Carta Marina des Olaus Magnus von 1539* (Göttingen: Gratia-Verlag, 1981), pp. 41ff: 'Der mehrfache Sinn der Carta Marina'.

19 Thevet, *Cosmographie universelle*, I (fol. 190v). On the circumstances of this miracle and the considerable echo it gave to Counter-Reformation propaganda, see François Secret, *L'Ésotérisme de Guy Le Fèvre de la Boderie* (Geneva: Droz, 1969), pp. 17–18.

20 Thevet, *Cosmographie universelle*, I, IX, 5 (fol. 294v).

21 Ibid.

22 Thevet, *Cosmographie universelle*, I, XII, 10 (fol. 441r): 'This too have I observed among the Savages of Antarctica: who, seeing us eat the stale and rancid fat that served us for capons and quails, reproved us by asking how it could be that we lived so long, given that such saltiness is quite harmful to the bodies of men.'

23 Ibid. (following the preceding passage).

24 Thevet, *Cosmographie universelle*, I, II, 8 (fol. 49r). The passage is used in refutation of the 'fibs' of Pomponius Mela, speaking of the lake of *Themyns* in Egypt, where had been seen 'an island floating on the water', 'of a marvellous size' and covered with 'countryside, waste land, woods, forests, and here and there fine cities'. Thevet's reference to '*my* Savages in Antarctic France' was thus intended to devalue the pseudo-science of the ancients.

The mention of the savages of Brazil, in its disqualifying function, is found again in the *Grand insulaire* (II, fols. 118v–119r), where it condemns the statements of Sebastian Münster concerning the fountains of youth, Cereus and Neleus, which the ancients situated on the island of Euboea: 'Which, if it were true, would have caused thousands of people, sick of wearing a white beard, to spare no amount of money to travel to these miraculous fountains, to get some water and polish their chins with it, and renew their beard ... Our Margageas, Perusians [*sic*] and Mexicans would take care not to visit such fountains, given that, as I have said elsewhere, they remove all the beard from their face ... But I pantagruelize too far here.'

25 I refer here to the fecund reflections of Hélène Clastres in 'Sauvages et Civilisés au XVIIIe siècle', in vol. 3 of François Châtelet (ed.), *Les Idéologies* (Paris: Hachette, 1978; Paris: Marabout Université, 1981), III, pp. 191–210.

26 On de Gérando's enterprise and its ideological foundations, see, apart from the reference in the preceding note, the study by Britta Rupp-Eisenreich: 'Christoph Meiners et Joseph-Marie de Gérando: un chapitre de comparatisme anthropologique', in D. Droixhe, and Pol-P. Gossiaux (eds), *L'Homme des Lumières et la découverte de l'autre* (Brussels: Éditions de l'Université de Bruxelles, 1985), pp. 21–47.

27 Paul Jove (Paolo Giovio), *Histoires* (Lyon: G. Rouillé, 1552), Books XVII, XVIII (pp. 277ff). This was Thevet's source for, notably, *Cosmographie universelle*, I, fols. 37–8 (the story of Selim's capture of Cairo in January 1517).

    On Sansovino's collection, see Stéphane Yérasimos, 'De la collection de voyages à l'histoire universelle: la *Historia universale de' Turchi* de Francesco Sansovino', *Turcica* 20 (1988), pp. 19–41.

28 On two of these 'colloquia', see David Dalby and P. E. H. Hair, 'Le Langaige de Guynee: a sixteenth century vocabulary from the Pepper Coast', *African Language Studies* 5 (1964), pp. 174–91; also, by the same authors, 'Le Langaige du Bresil: A Tupi Vocabulary of the 1540s', *Transactions of the Philological Society, 1966* (Oxford, 1967), 42–66.

29 According to Littré's dictionary (s.v. 'margajat'), which, for this figural meaning, gives two examples drawn from Boursault and Voltaire. But Thevet, the first to use the word in French (*Cosmographie universelle*, II, XXI, passim) is not mentioned.

30 Thevet, *Cosmographie universelle*. I, fol. 137r.

31 *Cosmographie universelle*, II, XXI, 14 (fol. 941r–v). Cf. Lussagnet, *Le Brésil*, p. 178.

32 *Cosmographie universelle*, II, XXI, 5, fol. 916v (Lussagnet, *Le Brésil*, p. 55). Cf. Thevet's later *Histoire de deux voyages aux Indes australes et occidentales* (fol. 47v), where this passage is again taken up.

33 Thevet, *Vrais Pourtraits et Vies des hommes illustres* (1584), II, Book VI, ch. 103 ('Robert Gaguin'), fols. 530r–532v. On the work of Robert Gaguin, see primarily Franco Simone, *Il Rinascimento francese, studi e ricerche* (Turin: Società Editrice Internazionale, Biblioteca di studi francesi 1, 1965), chs 2, 3, 4.

34 This was the jurisconsult Jason de Maino or Mayno (1435–1519). See Michel Reulos, *Comment transcrire et interpréter les références juridiques* (Geneva: Droz, 1985).

35 Thevet, *Vrais Pourtraits*, II, fol. 532v *in fine*.

36 See Lussagnet, *Le Brésil*, p. 40, n. 1. One could, of course, think here of another type of onomastic comparison: *maire* could, by a Latin etymology,

be linked to *maior*, 'greater', in the sense of an ancestor, elder or principal (cf. the dictionaries of Godefroy, Huguet, and the *FEW* of Walther von Wartburg, s.v. *maior*). But it is unclear whether Thevet used the term elsewhere with this added meaning of a substantive; whereas the Tupinamba *maira*, brought into French as *mair(e)*, was from the time of the *Cosmographie universelle* a familiar one. What reinforces my hypothesis is the association of this 'Mair' with discursive activity. From the (hi)stories of these 'good folks' to the 'mythistory' (*mythistoire*) of Jean Lemaire, the function of the mairs of Brazil and of France was enveloped by the generic term 'discourse' and, specifically, 'discourse of origins'.

37   *Cosmographie universelle*, II, XXI, chs 4–6 ('Du Cap de Frie, et legere creance des Sauvages dudit païs'; 'Institution du grand Caraibe, et des transformations faictes par leurs prophetes'; 'Poursuytte des transformations et croyance de ce peuple'), chs 3–5 in Lussagnet's edn.

38   See Hélène Clastres, *The Land Without Evil: Tupi-guarani Prophetism*, trans. Jacqueline Grenez-Brovender (Lanham, Md.: Rowman, 1975), pp. 51ff.

39   *Cosmographie universelle*, II, XXI, 6 (fol. 918r; cf. Thevet's *Histoire de deux voyages*, fol. 48v).

40   See ch. 2, above, nn. 54ff.

41   On the circumstances of this translation, see my work *Le Huguenot et le Sauvage* (Paris, 1990), ch. 3 and nn. 106, 108.

42   *The New Found Worlde, or Antarctike ... travailed and written in the French tong, by that excellent learned man, master Andrewe Thevet* (London: by Henrie Bynneman, for Thomas Hacket, 1568), fol. 2r ('Epistle'): 'To the right honorable Sir Henrie Sidney, Knight of the most Noble order of the Garter, Lorde President of Wales, and Marches of the same, Lord Deputie Generall of the Queenes Maiesties Realme of Ireland, Your humble Orator *Thomas Hacket* wisheth the favoure of God, long and happy life, encrease of honor, continuall health and felicitie.' The epistle begins as follows: 'None are more to be commended (right Honorable) than those who wer the first inventers and finders out of Artes and Sciences, wherwith mankind is beautified and adorned, without the which giftes he were but naked, barbarous and brutish, yea and a servile creature. It was not for nothing that the elders in times past did so muche celebrate the instituters of those things: as *Herodotus* writeth, that the Egiptians before all other men first found out the yere by the course of the Planets, and devided it into .xij. monthes. *Diodorus* assigned it to the *Thebanes*, the which standeth well with the opinion of *Herodotus*, bicause the *Thebanes* be a nation of Egipt. As *Numa* added to the yere Ianuary and February, *Romulus* ordred Marche, Aprill, and May, *Augustus* an other part, and so *Iulius Caesar* made up the perfect yere, as *Polidorus Vergilius* witnesseth in his boke *De inventoribus rerum*.'

On the project of Polydore Vergil, see Denys Hay, *Polydore Vergil. Renaissance Historian and Man of Letters* (Oxford: Clarendon Press, 1952), ch. 3 ('De inventoribus rerum'); also Jean Seznec, *The Survival of the Pagan Gods: The Mythological Tradition and its Place in Renaissance Humanism and Art*, trans. Barbara Sessions (Princeton: Princeton University Press, 1972), p. 22. Contrary to Seznec's view, however, it was not Polydore Vergil's Euhemerist excesses that caused his book 'On Inventors' to be put on the Index, but rather his Erasmianism. In fact, the augmented 1521 edition of that compilation vaunted the primitive Church and criticized the 'inventions' in the realm of the ecclesiastical establishment since the time of the

Apostles. It was this subversive content, tending in the same direction as the *Epitome topographica* of Vadianus, that is the point at issue in the fourth edition of the *Index de l'Université de Paris*, ed. J. M. De Bujanda, F. M. Higman and J. K. Farge (Sherbrooke and Geneva, 1985), 249, no. 243). Denys Hay (*Polydore Vergil*, p. 65) has rightly emphasized the vehemence with which, in dedicating the five books he added in 1521, Polydore Vergil denounced 'the jungle of Judaic observances spreading over the field of the Lord'. It was what led him to condemn the priests who encouraged people to venerate images of the Holy Family, and to inveigh against the trade in relics. But neither Héret (see below) nor Thevet, nor (later) Belleforest, seem to have been shocked by such audacities. See also n. 45.

43 *New Found Worlde*, fol. *3v: 'Thus (right honorable) we see, the valiant and curagious personages of the world have brought to passe many excellent enterprises, so that their name shall never dye, atcheved as well by sea as by lande, as this worthy traveller *Andrewe Thevit*, in this his Navigation of the New found World, which I have dedicated unto your honor . . .'

44 See Pierre-François Fournier, 'Un collaborateur de Thevet pour la rédaction des *Singularitez*', *CTHS. Bulletin de la Section de géographie* XXXV (1920), pp. 39–42; also my study, *André Thevet, cosmographe des derniers Valois*, ch. 4, nn. 50, 51. (Translator's note: Dares Phrygius was the hero of an apocryphal fifth-century work based on a character in the *Iliad*; it was popular in medieval times and translated into English in 1553 by Thomas Paynell.)

45 *New Found Worlde*, ch. 30 (fol. 47r). Cf. Polydore Vergil, *De rerum inventoribus*, Book 2, ch. 9 ('Who set furth bookes first, or made a library, Printing, paper, parchment. arte of memory'), fol. 53v: 'In memory excelled *Cyrus* kyng of Persye, whiche could call every man in his hoost by name. *Cyneas* the ambassadour of Pyrrhus the day after he came to Rome saluted every order of nobles by their proper names, *Mithridates* could speake twenty-two languages. *Julius Caesar* could wryte, reede, endite, and heare a tale al at ones, according to Spartian. Adrianus the emperour could do the same', *An Abridgement of the notable woorke of Polydore Virgile conteigning the devisers and firste finders out aswell of Artes, Ministeries, Feactes and civill ordinances, as of Rites, and Ceremonies, commonly used in the churche: and the originall beginning of the same*, trans. John Langley (London: Richard Grafton, 1546), fol. 47v.

46 *New Found Worlde*, ch. 31, fol. 48v.

47 Polydore Vergil, *Polydori Virgilii Urbinatis de inventoribus rerum libri tres* (Strasbourg: Schürer, 1515), Book 3, ch. 17 ('Quis primum instituerit artem meretriciam, aut invenerit tincturam capillorum, vel usum tondendi, & quando primum tonsores Romae'). Cf. Belleforest's trans. (p. 350: 'Plutarch in his Theseus'). A correction, here, to my 1983 edition of the *Singularitez*: on p. 61, n. 3, I indicate Plutarch's *Life of Theseus* (ch. 5) as the source for this passage. In fact, Polydore Vergil served as intermediary; and this was perhaps also the origin of the following misunderstanding on Thevet's part (loc. cit.): 'And in fact we find that Alexander king of Macedon commanded his people to take the Macedonians by the hair and beard, which they wore long; because at the time there were no barbers to cut or shave them.'

48 *New Found Worlde*, ch. 35 (fol. 55v): 'of the which ["Magike"] was the inventor as it is sayde Zamolxis and Zorastria, not he that is so common, but he that was sonne to Oromasia.' Cf. Polydore Vergil, *De rerum inventoribus*, I, 22 (in Belleforest's 1576 trans., p. 119).

49 *New Found Worlde*, ch. 35 (fol. 52v): 'These may be compared to Philon the first interpreter of dreames, and to Trogus Pompeius, that therein was very excellent. I might here bring in many things of dreames and divinations, and what dreams are true or no. Likewise of their kinds and the causes thereof, as we have been instructed by our elders ...' Cf. Pliny, *Natural History* VII.56. This chapter 'On the first Inventors of Many Things' was abundantly used by Polydore Vergil (see Belleforest's translation, I, 24, p. 127). (Translator's note: Hacket's translation of Thevet here makes no mention of Pliny or of Amphiction.)

50 *New Found Worlde*, ch. 38, fols. 57r–58v: '... being come forth, they assaile one another shoting of their arrowes: also with their Maces and Swords of wood, that to behold them it is a good passetime: they wil bite one another with their teeth in all places whereas they can take hold, chewing sometymes the bones of those whome they have vanquished and overcome beforetimes in the warrs, and eaten: to be short, they do the worst they can to feare and anger their enemies.' (Translator's note: Hacket did not translate a further detail in this passage: namely, that they chew each other also 'on the lips that they have a hole cut into'.)

This savage warfare, whose depiction seems to be from life itself, was in fact inspired by a model in Polydore Vergil (II, 10, fol. 64r–v in the 1544 edn): 'The ancients before the use of arms used to fight with fists, claws and by biting with their teeth: this was the beginning of battles. Then it became common to use stones and blows from sticks, according to Herodotus in his fourth book ... Diodorus affirms it in his first book, and that clubs and lion skins were the fighting gear of Hercules, given that arms were not found earlier, nor in his own time. Insults were avenged with great wooden clubs, and fighting men were covered and armed only with animal skins.'

From this fact, if one follows Polydore Vergil's 'lesson', it would seem that the Tupinamba combined in their manner of warfare several distinct stages of the arts of war.

The same ch. 38 of the *New Found Worlde* is again indebted to Polydore Vergil (fol. 70v) on the origin of treaties: 'and I think that if Theseus, the first author of a treaty with the Greeks, was present, he would be more hindered than he was then'; based on Plutarch's *Life* of Theseus, ch. XXXVII (via Polydore Vergil, Book 2).

51 *New Found Worlde*, ch. 42 (fols. 65r–66r) on the subject of the 'Cryb, a people of Thracia'; based on Polydore Vergil I, 4 (fol. 11 of the 1544 edn): 'On the birth of marriage, how peoples differ in this sacrament, and of those who join together carnally in public (in the manner of animals)'.

52 *New Found Worlde*, ch. 47 (fol. 74r), on the barter economy that was anterior to the use of gold coins. Based on Polydore Vergil, III, 16, or Pliny VII.56.

53 *New Found Worlde*, ch. 53 (fol. 83r); based on Polydore Vergil, III, 14 (fol. 124 of the 1544 edn): 'Ovid, however, says in the eighth of his *Metamorphoses* that Perdrix, the nephew of Daedalus, invented the saw on account of his sweat, following the example of the spines on the back of a fish, which have the form of a comb.' Cf. Ovid, *Metamorphoses* VIII.256–8. Again, the same page of Thevet borrows from Polydore Vergil (ibid.) traditions concerning the origin of 'ironworks' such as the 'hammer', 'saw', 'file' and 'nail'.

54 *New Found Worlde*, ch. 54 (fol. 84v). After Polydore Vergil (III, 7; fol. 111 of 1544 edn), who in turn cites Pliny (VII.56) and Vitruvius (II.1, on 'the life

of the first men' before the invention of fire and fixed living-places). See *Vitruvius on Architecture, from the Harleian manuscript 2767*, trans. Frank Granger, 2 vols (London: William Heinemann; New York: G. P. Putnam's Sons, 1931), vol. 1, pp. 77–81. In Polydore Vergil's words, 'Each man began to think of some remedy against the cold and ice, such that they made small edifices of daub with a rude earthen floor, held up with rods and switches tied together with vines and wicker.'

55 *New Found Worlde*, ch. 58 (fols. 90r–v, 92r), on the life of the first men and the origin of agriculture. Polydore Vergil (III, 2; fol. 99 of the 1544 edn) had already effected a modest synthesis on the question. It is to him that Thevet owes, notably (fol. 113v), this reference to Virgil (*Georgics* I.125): 'Never till Jupiter's reign did a farmer put fields under tillage . . .' *The Bucolics and Georgics of Virgil, rendered in English hexameters*, trans. Alexander Falconer Murison (London, New York, Toronto: Longmans, Green and Co., 1932), p. 53.

56 On Theseus, see nn. 47, 50ff, above; on Caesar, note 45. As for Lycurgus and Solon, the legislators of Sparta and Athens, their twin *Lives* recorded by Plutarch handed their antithetical wisdoms down to the men of the Renaissance. Thevet shows his debt to them in the *Cosmographie de Levant* (chs XXVII, XXVIII), and evokes the example of Solon in the *New Found Worlde* (ch. 51, fol. 80r–v) on the subject of restrictions put on the trade in honey; again, no doubt, on the authority of Polydore Vergil.

57 *Vrais Pourtraits*, II, VIII, chs 149, 150, 147, 145 (in the order mentioned).

58 All of these 'inventors' are mentioned in Book VII.56 of Pliny's *Natural History*.

59 On this filiation, see Denys Hay, *Polydore Vergil. Renaissance Historian and Man of Letters*, ch. 3, p. 58.

60 Ibid., pp. 60–1.

61 *New Found Worlde*, ch. 58 (fol. 93r). After Polydore Vergil, *De rerum inventoribus*, III, 2; fol. 99 of the 1544 edn): 'In the beginning, according to Pliny in his preamble to the sixteenth book of his natural histories, nature, the mother of all men and all things, did not teach men to live as delicately as they do now. For they were content to eat the fruits of the earth, produced and engendered without being cultivated, corrupted or violated with ironwork, as Ovid Nason says in the first of his Metamorphoses . . . Virgil says on this question in the first of his Georgics . . . cultivation was refined and added wheat and other grains when the forests were found to be sterile; even that of Dodona had begun to withdraw its fruit . . .' (Translator's note: Hacket's translation omits any mention of Ovid from the passage of Thevet's book in question here; a fact possibly related to its controversial nature, elaborated in n. 63.)

62 As du Bartas showed, in *Eden (Seconde Semaine)*, 'Premier Jour', vv. 271ff. See on this point my communication 'L'Art imite la Nature / La Nature imite l'Art: Dieu, du Bartas et l'Éden', *Actes du colloque Guillaume de Saluste du Bartas (Pau, 1986)* (Lyon: La Manufacture, 1988), pp. 167–84.

63 This lapse on Thevet's part (*New Found Worlde*, fol. 93r), evicting Cain in favour of Abel the shepherd, would arouse the sarcasm of Jean de Léry in his *Histoire d'un voyage* 3rd edn (Geneva, 1585), preface (fol. qqq³): 'He [Thevet] has published his ignorance, by saying that the holy Scripture mentions the farming of Abel; for if he puts on his spectacles he will find that he was a herder of sheep, and his brother Cain a cultivator of the soil.'

64 *New Found Worlde*, ch. 36 (fols. 53r–55v).
65 *New Found Worlde*, ch. 35 (fol. 53r). The conclusion of this chapter is
     as significant as it is vague: '... and they are truely idolaters, even as were
     the ancient Gentiles' [... ne plus ne moins que les anciens Gentils].
66 See on this point Lussagnet, *Le Brésil*, p. 39, n. 2 and p. 45, n. 1. The opinion
     regarding the deluge is evoked in the *New Found Worlde* in ch. 53 (fols. 82v–
     83r). The importance missionaries attached to this myth, in which they saw
     a prefigurement of Christianity and a proof of the unity of its Revelation,
     is well known.
67 *New Found Worlde*, ch. 58 (fol. 93r). On the invention of the cultivation of
     sweet potatoes and maize by the same Maire, see *Cosmographie universelle*,
     II, XXI, ch. 6 (fol. 918r) (Lussagnet, *Le Brésil*, pp. 61–2). Cf. also Thevet's
     Histoire de deux voyages aux Indes australes et occidentales (fol. 48r): Maire-
     Monan, appearing in the guise of a child and beaten by the Indians, 'made
     roots rain down on them, which they call *Yetic* and are like our rape; as well
     as flour, which they call *Avaty*'.
68 Cf. ch. XVI of Léry's *Histoire d'un voyage*.
69 Lussagnet, *Le Brésil*, pp. 39, 43–5, 66–72. Cf. Pierre Clastres, *Le Grand Parler.*
     Mythes et chants sacrés des Indiens Guarani (Paris: Seuil, 1974), pp. 95–9,
     reproducing, under the rubric 'Adventures of the Twins', a large extract
     from the *Cosmographie universelle* (II, fols. 919–20). It was previously
     transcribed by Lussagnet, *Le Brésil*, pp. 66–72.
70 See Claude Lévi-Strauss, *Structural Anthropology*, trans. Claire Jacobson and
     Brooke Grundfest Schoepf (London: Allen Lane, Penguin Press, 1968),
     p. 229.
71 *Cosmographie universelle*, II, fol. 918v (Lussagnet, *Le Brésil*, p. 65).
72 Ibid., fol. 920r (Lussagnet, *Le Brésil*, p. 71).
73 Gilles Deleuze, and Félix Guattari, *Rhizome* (Paris: Minuit, 1976), Intro-
     duction. Cf. p. 66: rhizomatic systems are defined by their opposition to
     arborescent (hierarchized) systems, and characterized furthermore by their
     openness to the exterior.
74 *New Found Worlde*, ch. 58 (fol. 93r).
75 Ibid., ch. 53 (fol. 82v).
76 Ibid. In the equivalent passage of *Cosmographie universelle* (XXII, 12, fol.
     937v; Lussagnet, *Le Brésil*, p. 161), Thevet does not reproduce this turn of
     phrase. (Translator's note: Hacket incorrectly translated Thevet's word
     'smoke' (*fumée*) in this passage, as 'sun'. This may, again, be associated with
     the controversy the work had aroused: cf. the following paragraph.)
77 Jean de Léry, *History of a Voyage ...*, pp. 166–7: 'However, I do not mean
     to say, much less do I believe or want you to believe, what a certain person
     has written: that is, that the savages of America, before this fire-making
     invention, dried their meat with smoke; for just as I hold that proverbial
     maxim of physics to be very true – that there is no fire without smoke – also,
     conversely, I think that he is not a good naturalist who would have us
     believe that there is no smoke without fire. By 'smoke' I mean – as did he
     of whom I am speaking – the kind that cooks meat. If, to get around this,
     he intended to say that he had heard about vapours and exhalations, he is
     making fools of us; for while we grant that some of these vapours are hot,
     they can by no means dry flesh or fish – they would instead make them
     moist and humid. Since this author, both in his *Cosmography* and elsewhere,
     complains so loudly and so often about those who do not speak just as he

likes about the matters he treats (but whose books he admits he has not thoroughly read), I entreat my readers to note the ludicrous passage I have cited about his preposterous hot smoke, which I herewith send straight back into his windbag of a brain.' Paul Gaffarel, in his edition of the *Singularitez* (Paris: Maisonneuve, 1878, p. 267, n. 1) makes no more sense than Léry of such a 'naïvety'.

78 Claude Lévi-Strauss, *The Raw and the Cooked* (New York: Harper and Row, 1969), 'Mythologics' pp. 7–12.

79 *New Found Worlde*, fol. 82v.

80 Ibid., ch. 61 (fol. 97v).

81 For an analysis of the 'mythic thought' at work in the description of Tupinamba cannibalism in the Renaissance, see my study 'Rage, fureur, folie cannibales: le Scythe et le Brésilien', in Jean Céard (ed.), *La Folie et le corps* (Paris: Presses de l'École Normale Supérieure, 1985), pp. 49–80, esp. 49–58.

82 Claude Lévi-Strauss, *The Savage Mind* (London: Weidenfeld and Nicolson, 1966), ch. 1, p. 20: 'One understands then how mythical thought can be capable of generalizing and so be scientific, even though it is still entangled in imagery.'

Chapter 4 Mythologics II: Amazons and Monarchs

1 Erwin Panofsky, *Studies in Iconology: Humanistic Themes in the Art of the Renaissance* (New York: Harper and Row, Icon Editions, 1972), ch. 3.

2 Thevet, *Vrais Pourtraits et Vies des hommes illustres* (1584), vol. 2, Book VIII, ch. 149, fol. 661r ('Quoniambec').

3 Guillaume de Salluste du Bartas, *Works of Du Bartas.* trans. Urban T. Holmes (Chapel Hill: University of North Carolina Press, 1940), vol. 3, p. 154; or see *Bartas: His Devine Weekes & Workes* (London, 1605; repr. Delmar, N.Y.: Scholarly Facsimiles, n.d.), Second Week, Second Day ('Colonies'), vv. 275–8.

4 See, on this point, the contribution of Huguette Zavala, 'L'Allégorie de l'Amérique au XVIe siècle', in *La Renaissance et le Nouveau Monde*, ed. Alain Parent, (Quebec: Musée de Québec, 1984), pp. 129–47. Cf. also Jean-Claude Margolin, 'L'Europe dans le miroir du Nouveau Monde', in *La Conscience européenne au XVe et au XVIe siècle* (Paris: Presses de l'École Normale Supérieure, Collection de l'ENSJF, 1982), pp. 235–64.

5 Thevet, *Singularitez.* fol. 71v; *Cosmographie universelle*, II, XXI, fol. 942v. (Translator's note: these plates were not reproduced in Hacket's translation, *The New Found Worlde*.)

6 *Singularitez*, ch. 38, fol. 71v. (Translator's note: this is my translation of the passage, also cited in ch. 3, n. 50; Hacket fails, as noted, to mention the perforated lips.)

7 *Cosmographie universelle*, II, XXI, fol. 942v (Lussagnet, *Le Brésil*, p. 183).

8 On the distinction between these styles, see my study 'Les représentations du sauvage dans l'iconographie relative aux ouvrages du cosmographe André Thevet', *Bibliothèque d'Humanisme et Renaissance*, XL (1978), pp. 583–95.

9 Polydore Vergil, *De rerum inventoribus*, II, 10 (fol. 64r–v in the 1544 edn). Cf. ch. 3, n. 50 above.

10 Étienne Delaune, *Combats et Triomphes*, a set of twelve engravings forming a frieze with dark background, 65–7 mm × 218–22 mm (BN; Dept.

Estampes: Ed 4 pet. fol.). It is described in Robert-Dumesnil, *Le Peintre-graveur français*, IX, pp. 87, 281–92. Cf. also André Linzeler, *Inventaire du fonds français. Graveurs du seizième siècle*, vol. 1: *Androuet du Cerceau-Leu* (Paris: M. Le Garrec, 1932), pp. 272–5 (nos 275–86).

11 For an overview of the artefacts of ancient Brazil that had reached Europe, see the study of Christian Feest, 'Mexico and South America in the European *Wunderkammer*', in Oliver Impey and Arthur MacGregor (eds), *The Origins of Museums: The Cabinet of Curiosities in Sixteenth- and Seventeenth-Century Europe* (Oxford: Clarendon Press, 1985), pp. 237–46.

12 On these ornaments, see Alfred Métraux, *La Civilisation matérielle des tribus Tupi-Guarani* (Paris: Paul Geuthner, 1928), pp. 137– 8 (on diadems), 148 (on ostrich feather boas).

13 *Singularitez*, fol. 83r; *Cosmographie universelle*, II, XXI, fol. 927v (Lussagnet, *Le Brésil*, p. 106). Cf. the verbal description in *The New Found Worlde*, fol. 69r. On the Tupinamba's cloaks of red ibis feathers, see Alfred Métraux, 'A propos de deux objets tupinamba du musée du Trocadéro', *Bulletin du Musée d'ethnographie du Trocadéro* (January 1932), pp. 3–12.

14 On this tradition, see Claude Gaignebet and Jean-Dominique Lajoux, *Art profane et religion populaire au Moyen Age* (Paris: Presses Universitaires de France, 1985), pp. 79–87 ('L'oncle des bois') and passim. One of the late illustrations of the killing of a prisoner during ritual banquets of the Tupinamba shows the victim metamorphosed into a bear, being struck with a blow from the club of a 'feathered' savage rigged out with angel's wings. This retouched engraving was intended to illustrate the joint edition of the accounts of Hans Staden and Jean de Léry, *Historia Antipodum Order Neue Welt* (Frankfurt: Matthias Merian and Johann Ludwig Gottfrid, 1631), vol. III (BN: Rés. G. 431–3).

15 Cf. *Cosmographie universelle*, II, fol. 942v (Lussagnet, *Le Brésil*, p. 183): 'And when they have taken someone prisoner, they stick their finger into his hare-lip (which they all have, and in which they keep their precious stones), and display their magnificence and bravado by pulling him about.'

16 Ibid.

17 Delaune, *Combats et Triomphes*, plate 4. It is reproduced in Rabelais, *Oeuvres complètes*, pp. 146–7.

18 See Panofsky, *Studies in Iconology*, ch. 2 ('The Early History of Man in Two Cycles of Paintings by Piero di Cosimo').

19 Antoine Jacquard, *Les Divers pourtraicts / et figures faictes sus les meurs / Des habitans du Nouveau Monde / Dedié / A Jean le Roy Escuyer Sieur de la / Boissiére gentilhomme poictevin / Cherisseur des Muses* (a set of thirteen numbered pieces, incl. frontispiece and twelve plates arranged as friezes, each enclosing four subjects in arcades: BN: Estampes, Of. 5/4°), dated *c*.1620. The bibliographical description given by Roger-Armand Weigert, in *Inventaire du fonds français*, p. 445 (nos 35–47), is extremely brief.

For the few biographical facts known about Antoine Jacquard – a Poitiers goldsmith, engraver and possibly surveyor, whose work extends from 1613 to 1640 – see the contributions of Henri Clouzot in *Antoine Jacquard et les graveurs poitevins au XVIIe siècle* (Paris: Librairie Henri Leclerc, 1906); and of Dr Édouard-Théodore Hamy, 'L'*Album des habitans du Nouveau Monde*', *Journal de la Société des Américanistes de Paris*, n. s. IV, no. 2 (1908) (BN: 4° P. 1213).

20 This Jean Le Roy, sieur de la Boissière, to whom the album is dedicated, was from a family of notables of whom three had been mayors of Poitiers

(in 1293, 1482, 1559). The same Jean Le Roy is mentioned in eulogizing terms in *Le Jardin, et Cabinet Poetique de Paul Contant apoticaire de Poictiers* (Poitiers: Antoine Mesnier, 1609), a work dedicated to Sully (BN: S. 3726 and S. 4044); Antoine Jacquard may have also illustrated this work, which is fertile in marvels and nourished by a reading of the *Seconde Sepmaine* of du Bartas (notably on pp. 64–7, where the *Imposture* is cited in connection with the satanic dragon). Paul Contant's *Jardin et Cabinet* contains an appreciable number of Americana: the toucan (10, p. 68, plate VII); the 'light canoe' (11, ibid.); the 'Espadon de mer' (swordfish) (13, p. 69, plate VII); a 'strange lizard' that may refer to the iguana (14, p. 69, plate IV); the tattoo and the armadillo (15 and 16, p. 70, plate II); the 'bat' or vampire (18, p. 71, plate VII); or the feathered maracas (39, p. 84, plate VII) poetically described as an 'American fruit that those idolatrous people adore like a God of gold, silver or plaster – by superstition!; that the brutal hand of the cruel Carib makes resonate with such a proud sound when he puts into these worthy fruits some grain [maize] of his land or small pebbles. It is decked out with the finest feathers of the toucan, arat, or other birds of the greatest rarity, and with it the ministers of Satan, dressed in similar fashion, go from door to door and village to village, and around their houses. These fruits, so dressed up, they plant and harvest, with orders to the fathers of families to give them everything for their alimentation; for the fruit Maracas is a god, who feeds only at night.' (Translator's note: here translated as prose.)

Maracas are found in several places in Antoine Jacquard's Brazilian gallery: frontisp., 1; plate 4, 1–4; plate 6, 1–3; plate 7, 3 and 4; plate 8, 1 and 4; plate 9, 4; plate 10; plate 11, 4.

Dr E.-T. Hamy ('Album des habitans', p. 13) thought he saw an allusion to Thevet in the following eulogy, introducing the description of the vampire in Contant's *Jardin et Cabinet Poetique* (18, p. 71, plate VII) (translated as prose): 'A faithful writer whose authentic plume has revealed to the French that other, Antarctic, France and told us of strange and distant lands, their fashions and manners and the rarest beauties of the Americans, tells a history as fine to hear as it is difficult to believe ... But the fidelity of that great person renders the latter account in all places as if it were that of an eye-witness, who had seen for himself the bright blood flowing from his big toe.'

21 The fourteen volumes of Théodore de Bry's *America* were published in Frankfurt from 1590 to 1634. On this collection, abundantly illustrated and imitated for two centuries, see Michèle Duchet et al., *L'Amérique de Théodore de Bry. Une collection de voyages protestante du XVIe siècle* (Paris: Éditions du CNRS, 1987).

22 De Bry, *Americae tertia pars* (Frankfurt, 1592), pp. 71, 124–6 and passim. Cf. Duchet et al. *L'Amérique*, pp. 117, 183–5.

23 I note here the ethnographic remarks of Dr E.-T. Hamy ('Album des habitans', p. 16), who sees the axe with an extremely curved end that accompanies the second figure of plate 6 as a club of 'Antillian form'. But one thinks also of a type of axe in the form of an anchor that was characteristic of the Gê of eastern Brazil, which can be seen in the collection of the Château d'Ambras, now in the Museum für Völkerkunde, Vienna. On this family of objects and their provenance, see Feest, 'Mexico and South America', p. 242 and plate 91.

24 Anon, *Histoire naturelle des Indes: contenant Les Arbres, Plantes, Fruits,*

Animaux, Coquillages, Reptiles, Insectes, Oyseaux, etc. qui se trouvent dans les Indes; representés Par des Figures peintes en couleur naturelle; comme aussi les diférentes maniéres de vivre des Indiens; savoir: La Chasse, la Pêche, etc. Avec Des Explications historiques. MS, c.1590 (Morgan Library, N.Y.: MS. 3900), fol. 85r ('HINDES DE IHONA'). Here is the text that accompanies this watercolour: 'When these Indians are masters of their / enemies, they make them lie on the ground and thrash them, and then give them a blow / with their sword on the head and, as the blood / begins to flow, collect it hastily, considering / by this means to make the body better for roasting / so as to eat it in great solemnity; thereby / gaining great prestige.'

25 André Chastel, *La Crise de la Renaissance* (Geneva: Skira, 1968), pp. 146–7). Cf. also Gisèle Mathieu-Castellani, *Emblèmes de la mort. Le dialogue de l'image et du texte* (Paris: Nizet, 1988), pp. 121–30.

26 Thevet, *New Found Worlde*, ch. 63, fol. 103r. For a diagonal reading of Thevet's amazonian episodes, see Anna-Luigia Villani, 'La leggenda delle Amazzoni', *Bollettino dell'Istituto di Lingue Estere (Genova)* 13 (1983), pp. 93–112, esp. 95–9. This author bases herself on the thesis of Giuliano Gliozzi, in *Adamo e il nuovo mondo* (Florence: La Nuova Italia, 1977).

27 *New Found Worlde*, fol. 101r.

28 Ibid., fol. 102r.

29 Ibid., fol. 102v. The engraving was reused, as we shall see, in the *Cosmographie universelle* (II, XXII, ch. 3, fol. 960v).

30 All references here are to Georg Friederici, *Die Amazonen Amerikas* (Leipzig: Simmel, 1910), passim. Cf. also Juan Gil, *Mitos y utopías del Descubrimiento, 3: El Dorado* (Madrid: Alianza Universidad, 1989), pp. 195ff.

31 Anglerius (Peter Martyr d'Anghiera), *De Orbe Novo: The Eight Decades of Peter Martyr d'Anghiera*, trans. Francis A. MacNutt, 2 vols (New York: Burt Franklin, 1971), Fourth Decade. (Translator's note: the original translation, *The Decades of the newe Worlde of West India Conteyning the navigations and conquestes of the Spanyardes* ... (London: Will. Powell, 1555), included only the first three decades.)

32 See my article, 'Le nom des Cannibales, de Christophe Colomb à Michel de Montaigne', *BSAM*, 6th series, nos 17–18 (June 1984), pp. 51–74.

33 These are myths no. M. 156 (Apinayé: the seducing cayman), M. 287 (Carib: the seducing jaguar) and M. 327 (Warrau: the origin of tobacco), recorded by Claude Lévi-Strauss in *From Honey to Ashes: Introduction to a Science of Mythology* (vol. 2), trans. John and Doreen Weightman (Chicago: University of Chicago, 1983).

34 C. Rhodiginus, *Lectiones Antiquae*, IX, 12 ('Item de Amazonibus scitu dignum'), p. 461 E of the Geneva edn (Philippe Aubert, 1620); N. Perotti, *Cornucopiae seu Latinae linguae commentarii locupletissimi* (Basle: apud V. Curionem, 1526) ('Amazones'); A. Calepino, *Dictionarium, quarto et postremo ex R. Stephani Latinae linguae Thesauro auctum* (Paris: R. Estienne, 1553) ('Amazones'); R. Estienne, *Dictionarium propriorum nominum vivorum, mulierum, populorum, idolorum, urbium [etc.]* (Paris: by the author, 1541) ('Amazones'). Thevet had already used the same lexicographic instruments for his *Cosmographie universelle*. See above: ch. 2, nn. 11ff.

35 *New Found Worlde*, ch. 63 (fol. 102v).

36 Ibid.

37 M. 327, in Lévi-Strauss's classification (see n. 33).

38 *New Found Worlde*, fol. 101r: 'these *Amazones* of which we speake, are

retired, inhabiting in certaine Islands which are to them as strong holdes, having alwayes perpetuall warre with certaine people ...'

39 *New Found Worlde*, fol. 103r–v (incorrectly numbered 74).
40 François Delpech, 'La légende de la Tierra de Jauja dans ses contextes historique, folklorique et littéraire', *Texte et Contexte (XVe Congrès de la Société des Hispanistes français)*, in *Trames* (special issue) (Limoges, 1980), pp. 79–98, esp. 95.
41 *New Found Worlde*, ch. 38 (fol. 58r).
42 Ibid., ch. 42 (fol. 64v).
43 Thevet, *Singularitez*, fol. 101r. (Translator's note: the plate is not in *The New Found Worlde*.)
44 Lévi-Strauss, *Tristes Tropiques*, ch. 27.
45 Thevet, *Cosmographie universelle*, II, XXI, ch. 14 (fol. 941v).
46 François Hartog, *The Mirror of Herodotus*, p. 216. On the amazons of antiquity, cf. Jeannie Carlier-Détienne, 'Les Amazones font la guerre et l'amour', *L'Ethnographie*, 74, 1 (1980), pp. 11–33.
47 *Cosmographie universelle*, vol. I, XII, chap. 12 (fols. 443v–446r). This chapter conjugates with the description of the dual archipelago of Imaugle and Inébile an evocation of the legendary magnetic mountains. It can be compared to a passage in the *Isolario* of Benedetto Bordone, 2nd edn (Venice: Nicolo d'Aristotile detto Zoppino, 1534) (BN: Rés. J. 169), fol. LXXr–v, which had already presented the same association and presented it, furthermore, in the form of a map.
This Indian Ocean island of amazons, with its male appendage of Inébile (destined, for its part, to ensure the continuity of procreation), appears for the first time in Western literature with Marco Polo: see *Le Devisement du monde* (Book 3, ch. CXC: 'Ci devise des îles Mâle et Femelle'), ed. Stéphane Yérasimos (Paris: La Découverte, 1980, p. 474). Its location approximately corresponded to the Islands of Kuria Muria off the coast of Oman. Thevet thus displaces the archipelago eastwards and, as we shall see, gives a new mythical variant of it, in so far as fecundation now takes place on the men's island, and not on that of the women as in a tradition common to most 'Amazonias'. (Translator's note: *The Book of Ser Marco Polo the Venetian, Concerning the Kingdoms and Mervels of the East*, trans. and ed. Henry Yule (London: John Murray, 1903, pp. 404–6) gives four possible referents: Socotra, or islands off Somalia or Malaysia or in the China Sea. The latter had three females to every five males, who hunted using the bow and arrow.)
After Marco Polo, the world map of Fra Mauro, dated between 1457 and 1459, mentioned twin islands of men and women beyond Cap Diab (the Cape of Good Hope) in a westerly direction; which had the effect of making Amazonia enter into the Atlantic region. From all of this Alexander von Humboldt would conclude that 'the fiction of amazons has penetrated throughout all the zones; it belongs in the narrow and uniform circle of reveries and ideas, in which the poetic or religious imagination of all races of men and of all epochs moves almost instinctively', (*Examen critique de l'histoire de la géographie du Nouveau Continent et des progrès de l'astronomie nautique aux quinzième et seizième siècles*, Paris: Gide, 1836–9, vol. I, p. 336. (Translator's note: this work was published in Berlin (*Kritische Untersuchungen . . .*, trans. J. L. Ideler) in 1836, but there is no English version.)
48 Guillaume Postel, *Les Tres-merveilleuses Victoires des Femmes du Nouveau*

Mode, et comment elles doibvent à tout le monde par raison commander (Paris: Jean Ruelle, 1553; reissued in the eighteenth century) (BN: Rés. D²10159). Cf. the title of the first chapter, 'Des Admirables Excellences et faictz du sexe feminin, et comment il faut qu'il domine tout le monde' (On the admirable qualities and facts of the feminine sex, and why it must come to dominate the world). (Translator's note: this work is in the British Library in another 1553 edn, by I. Gueullart, but is untranslated into English.)

49   This cosmological system is described in another work by Guillaume Postel, *Des merveilles du monde, et principalement des admirables choses des Indes et du Nouveau Monde* (Paris [?], 1553) (BN: Rés. D²5267). On this work, see my study 'Cosmologie et *mirabilia* à la Renaissance: l'exemple de Guillaume Postel', *Journal of Medieval and Renaissance Studies* 16, no. 2 (Fall 1986), pp. 253–79.

50   Guillaume Postel, *Des merveilles du monde*, ch. XXVI, fol. 92r.

51   Postel, *Les Tres-merveilleuses Victoires*, ch. 1, p. 3. Thevet takes Postel up on this precise point; '... although he who made the booklet entitled "The Most-Marvellous Victories of the Women of the World" gives us to understand that they live even today in southern Africa and the lands of Peru, the ones who were formerly so celebrated among the Greeks and Latins' (*Cosmographie universelle*, I, fol. 444r).

52   *Cosmographie universelle*, I, fol. 225r. The headliner of this passage, used in refutation of the ancient fables about amazons, is explicit: 'The Most-false History of Amazons'. The cosmographer did not think it useful to pronounce his own self-critique on this matter, as he would later do in a chapter on the amazons of Brazil (ibid., II, fol. 960r; cf. below, nn. 57ff). But he did not hesitate to formulate this very general and, when applied to his own case, rather rash maxim: 'Thus I say that any man of good judgement should note how much difference there is between history and fable; and that history, whether that of the ancients or of the moderns, must follow that path of truth.'

53   Ibid. I, fol. 444r.

54   Ibid. I, fol. 444v.

55   *Cosmographie universelle*, I, fol. 445v. The theme of feminine cruelty, already present in the *New Found Worlde* (ch. 63, fol. 102v), strongly tempers Thevet's eulogy of the valiant warrior women.

56   *Cosmographie universelle*, fol. 445v. On the fantasmatic complex associating islands with eroticism, one might consult with caution the suggestive study of Abraham A. Moles, 'Nissonologie ou science des îles', *L'Espace géographique* 4 (1982), pp. 281–9, esp. 286–7.
    The association of insular structures with amazons is again found – in the mode, it is true, of negation – in a chapter of Thevet's *Grand Insulaire* devoted to the islet of 'Strongile' in the Aegean Sea (vol. II, fol. 103v; BN: Ms. fr. 15453). The 'Rock of Penthesilea', a dangerous reef and scene of shipwrecks, provokes fertile conjectures on the part of Thevet, who ends up rejecting the 'quite fabulous' history of amazons.

57   *Cosmographie universelle*, II, XXII, ch. 3 (fol. 960r).

58   *Cosmographie de Levant* (1556), ch. 37, p. 136, line 7.

59   *Cosmographie universelle*, II, fol. 960r.

60   This positivistic type of explanation consists of seeking beneath the legend for an established social fact pointing in the same direction as its mythical

transposition. It is the causal schema retained by Georg Friederici, *Die Amazonen*, and by Enrique de Gandía in *Historia crítica de los mitos y leyendas de la Conquista americana* (Buenos Aires: Centro Difusor del Libro, 1946), ch. 6, pp. 75–87). The latter critic sees in the multitude of New World amazons a hyperbolic and distorted reflection of the Virgins of the Sun at Cuzco. Such a reductive schema is again found, paradoxically, in the work of Claude Lévi-Strauss, in a study devoted to the anthropology of the family: the young concubines of the Nambikwara chief, who prefer the military calling and the company of warriors to domestic work, could have provided the model for the legendary amazons of Brazil. See Claude Lévi-Strauss, *The View from Afar*, trans. (from the German edn) by Joachim Neugroschel and Phoebe Hoss (London: Blackwell, 1985; New York: Basic Books, 1987), pp. 51–2. This is a manifestly less formal analysis than his reconstruction in *From Honey to Ashes*, mentioned above.

61   Thevet, *Cosmographie universelle*, II, fol. 960r.
62   It is quite remarkable to see the same type of reasoning taken up by Thevet in his *Vrais Pourtraits et Vies des hommes illustres* (2, IV, ch. 25, fol. 280r), in order to exonerate Joan of Arc of the accusation brought against her by the English. If she donned for battle the 'dress of a man', it was not simply to behave like an amazon (as was implied in the reproach), and thereby to contravene the laws of nature. In fact, Thevet would have her ranked among those women who 'with a stout and virtuous heart with which to resist the enemy or defend their homeland, have quit their feminine costume to put on the arms and garb of war'. There follow the examples of 'Semiramis, queen of the Assyrians'; 'Thomiris, princess of the Shiites'; of Artemisia, Camilla, Lesbia, and Maria, queen of Hungary. Thevet could have added to this suite of *exempla* the valiant islanders of Imaugle and the diligent spouses of the Margageas. It is true that, in the case of Joan of Arc, there was the essential added ingredient of miracle, God having, by her 'frail' means, restored to France her liberty. But if Thevet does not mention the amazons in his chapter on 'Jeanne la Pucelle', Postel, on the other hand, in the *Tres-merveilleuses victoires*, makes the 'good woman of Lorraine' an intermediate step between the ancient warrior women and the new Eve, Mother Johanna of Venice (chs VI and VII).
63   *Cosmographie universelle*, II, fol. 960v.
64   Ibid.
65   See my study 'Fortunes de la singularité à la Renaissance: le genre de l'Isolario', in *Studi francesi* 84 (Sept.–Dec. 1984), pp. 423–7.
66   Jean de Léry, *Histoire d'un voyage faict en la terre du Brésil* (3rd edn, 1585), ch. 3, p. 33.
67   Jean de Léry, *Historia Navigationis in Brasiliam, quae et America dicitur* (Geneva: Eustache Vignon, 1586), 'Autores in hac historia citati' (at end of table of contents). This was a translation of the 1585 French edition.
68   Jean de Léry, *Histoire d'un voyage* (1585), ch. III, p. 35.
69   Among avatars of the myth of amazons after Thevet and Léry, one can note the stages constituted by the voyage accounts of Yves d'Evreux (*Suitte de l'Histoire des choses plus memorables advenues en Maragnan es annees 1613 et 1614. Second traité* (Paris: François Huby, 1615), ch. VIII, fols. 23v–24v), and of Jean Mocquet: *Voyages en Afrique, Asie, Indes Orientales et Occidentales. Faits par J. M., Garde du Cabinet des singularitez du Roy, aux Tuilleries* (Rouen: Jacques Caillové, 1645; 1st edn 1617), p. 102. The

prudent reservations of the former – a Capuchin friar who emphasizes the inadequacy of the legend of amazons to account for Indian realities – contrast with the fantasmatic delirium of Mocquet, who saw the wild, bow-wielding women as incestuous mothers and sisters: for in effect, since they did not keep the males they engendered but sent them back to their putative fathers, there was 'every appearance that the sons they had given over to those Indians could later have relations with their sisters and close relatives'.

70 On Hans Staden, see the recent contribution of Elisabeth Luchesi, 'Von den "Wilden, Nacketen, Grimmigen Mensch-fresser Leuthen, in der Newenwelt America gelegen". Hans Staden und die Popularität der "Kannibalen" im 16. Jahrhundert', in the catalogue of the exhibition *Mythen der Neuen Welt. Zur Entdeckungsgeschichte Lateinamerikas* (Berlin: Frölich und Kaufmann, 1982), pp. 71–4.

71 Hans Staden, *Wahrhaftige Historia*, trans. Henri Ternaux-Compans (Paris: A.-M. Métaillé, 1979), pp. 88–9. Cf. Thevet, *Cosmographie universelle*, II, fol. 924r–v. See, on the latter passage, Suzanne Lussagnet's note in her edition of André Thevet, *Le Brésil*, pp. 88–9, n. 3.

72 *Cosmographie universelle*, II, fol. 924r (Lussagnet, *Le Brésil*, p. 89).

73 Ibid., fol. 924v (Lussagnet, *Le Brésil*, p. 92). This scene is already present in an embryonic state in the *Singularitez* of 1557 (fols. 103v–104r; *New Found Worlde*, fol. 84r). But there Quoniambec, 'the most known and re-nowned King of the entire country', as yet makes a pale figure, by compari-son with his later avatars in the *Cosmographie* and *Vrais Pourtraits*.

74 *Cosmographie universelle*, II, fol. 924r (Lussagnet, *Le Brésil*, p. 89).

75 Ibid., fol. 952v (Lussagnet, *Le Brésil*, p. 232). See the plate entitled 'Ruse de Quoniambech'.

76 Ibid., fol. 924r (Lussagnet, *Le Brésil*, p. 92): 'This is how you would have seen his palace, which was a house as rich as that of others, all adorned and decorated outside with the heads of the enemies he had massacred and eaten.' The hyperbolic term 'palace' had appeared already in the *Singularitez* (fol. 103v; see *New Found Worlde*, fol. 84r).

77 *Cosmographie universelle*, II, fol. 924r (Lussagnet, *Le Brésil*, p. 92).

78 Ibid., fol. 924v (Lussagnet, *Le Brésil*, p. 93).

79 Léry, *Histoire d'un voyage* (3rd edn, 1585), pp. 146–7.

80 Ibid., Preface. Fol. B7v of the 2nd edn (Geneva: Antoine Chuppin, 1580).

81 Thevet, *Vrais Pourtraits et Vies*, vol. II, Book VIII, chs 135 ('Jules Cesar, Premier Empereur') and 138 ('Tamerlan, empereur des Tartares').

82 Ibid., ch. 149 (fol. 661v).

83 Cf. Montaigne, *Essays* (ed. Screech), I, 31 (p. 240): the Brazilians Montaigne met in Rouen 'said (probably referring to the Swiss Guard) that they found it very odd that all those full-grown bearded men, strong and bearing arms in the King's entourage, should consent to obey a boy rather than choosing one of themselves as a Commander...' The paradoxical thought of Montaigne joins here that of La Boétie, in his *Discours de la servitude volontaire*.

84 Jean de Meung, *Le Roman de la Rose* (thirteenth century, ed. Félix Lecoy), vv. 9579–82. See Guillaume de Lorris and Jean de Meun, *Romance of the Rose*, trans. Harry W. Robbins (New York: Dutton, 1962), p. 193. (Trans-lator's note: a prose translation, closer to the original, is that of Charles Dahlberg in *The Romance of the Rose* (Princeton: Princeton University Press,

1971), p. 172: 'They elected a great scoundrel among them, the one who was largest, with the strongest back and limbs, and made him their prince and lord.') These lines were cited by Lescarbot in *Histoire de la Nouvelle France*, 3rd edn (Paris: Adrian Périer, 1617), Book VI, ch. 24, p. 944. See *Lescarbot: The History of New France* (Toronto: Publications of the Champlain Society, 1914), p. 265; or the 1609 translation by Pierre E. Erondelle, *Nova Francia, or, the Description of that Part of New France, Which is One Continent with Virginia* (reprinted 1977 by Walter J. Johnson Press).

85 *Cosmographie universelle*, II, fol. 924r; *Vrais Pourtraits*, II, VIII, 149, fol. 661r. These two engraved portraits are both reproduced by Suzanne Lussagnet, *Le Brésil*, pp. 90–1; cf. n. 71.

86 Alfred Métraux, *La Civilisation matérielle des tribus Tupi-Guarani* (Paris: Libraire orientaliste Paul Geuthner, 1928), ch. 34, p. 266 ('Insignes de commandement').

87 Lescarbot, *History of New France*, p. 265, citing ch. VII of the *Germania* of Tacitus: 'The power of their kings is not unlimited or infinite, but they guide the people rather by example than by commandment.' Lescarbot enunciates, however, the following distinctions: 'In Virginia and Florida they are more honoured than among the Souriquois [or Micmac]; but in Brazil they take for captain him who has taken and killed the most prisoners, without his children being able to inherit that dignity.'

88 Thevet, *Vrais Pourtraits*, vol. II, Book VIII, ch. 149 (fol. 661v).

89 Léry, *Histoire d'un voyage* (3rd edn, 1585), Preface (fol. qq 5).

90 Ibid., fol. qqq 5.

91 Lescarbot, *History of New France*, p. 265; cf. n. 87, above.

92 Pierre Clastres, *Society Against the State: Essays in Political Anthropology*, trans. Robert Hurley and Abe Stein (New York: Zone, 1988), ch. 2 ('Exchange and Power: Philosophy of the Indian Chieftainship').

93 Thevet, *Histoire de deux voyages*, fols. 32v–33r. It is true that on this page our cosmographer and 'universal insulist' is referring to 'the Isle and River of St Dominique' in the latitude of 'Fernambourg' (Pernambuco), not of riverine tribes of the Rio de Janeiro seaboard.

94 See on this point *The Tupinamba*, in the *Handbook of South American Indians* (Washington: The Smithsonian Institution, 1948), vol. 3, pp. 113–14; also (apart from the remarks mentioned of Alfred Métraux) Lussagnet, *Le Brésil*, p. 252, n. 2.

95 Concerning, for example, the chief's oratorical privileges, Pierre Clastres writes in *Society Against the State* (p. 175): 'in no circumstance does the tribe allow the chief to go beyond that technical limit; it never allows a technical superiority to change into a political authority. The chief is there to serve society; it is society as such – the real locus of power – that exercises its authority over the chief. That is why it is impossible for the chief to reverse that relationship for his own ends, to put society in his service, to exercise what is termed power over the tribe: primitive society would never tolerate having a chief transform himself into a despot.' It is evident that André Thevet could only give a diametrically opposed interpretation of Quoniambec's ostentatious 'gift of the gab'. In his eyes, on the contrary, it was the efficacious and terrifying word of a monarch whose power was not only political and military, but was exerted magically over men and events.

96   Montaigne, *Essays*, III, 6 ('On Coaches'). The best interpretation of this
     chapter is that of Marcel Bataillon: 'Montaigne et les conquérants d'or',
     Studi francesi, ann. II, fasc. III (September–October 1959), pp. 353–67.
97   Thevet, *Vrais Pourtraits*, II, VIII, 149 (fol. 661r).
98   Thevet, *Cosmographie universelle*, II, fol. 906v. Cf. the *Grand Insulaire*, I,
     fol. 269r, where the same Thevet amplifies the 'giantist' digression in his
     earlier work: 'And because many people scruple to believe what they have
     not been able to see on their own doorstep, I am very pleased to give an
     account here of some giants that have been seen and recognized in our own
     France, and in so doing, to keep from falling into the ditch of incredulity
     those poor unbelievers who claim to have heard word of the existence of
     giants today. At the cloisters of the Jacobins of Valence one still sees the
     portrait of a giant named Buard, who was eight cubits high, together with
     the bones of this monstrous man. At Loches, Captain Pontbriant and some
     others found in an underground chamber a seated man of marvellous stat-
     ure, who from the proportions of his bones was thought to have been some
     eight feet tall. I will say nothing of the Giant of Saint-Germain-des-Prés at
     Paris; nor of the one said to have been defeated by Arthur, king of the
     Albionic Britons, on the Isle of Notre–Dame in this same city of Paris; nor
     of several others, so as to avoid prolixity.' For a recent synthesis on the
     question, see Jean Céard, 'La querelle des géants et la jeunesse du monde',
     The Journal of Medieval and Renaissance Studies 8, no. 1 (Spring 1978),
     pp. 37–76.
99   Thevet, *Grand Insulaire*, I, fol. 269r: 'Those who consider that the great
     Deluge destroyed in any way the race born of the earth, will find from the
     discourse I am about to present how much they are mistaken.'
100 Ibid.
101 *Cosmographie universelle*, II, fol. 906v. This chapter, entitled 'Of an island
     where the men are ten to twelve feet high' (fols. 903–6), is a redundant and
     hyperbolic transformation of the sober narration of Antonio Pigafetta,
     Magellan's companion on his voyage.
102 Girolamo Cardano, *De Rerum Varietate libri XVII*, Book 1, ch. IV, p. 22.
     Cited from the translation of Jean Céard (*La Nature et les prodiges*, p. 232).
103 For an evocation of the different *scenarios* of the Magellan interview, see
     my communication 'La flèche du Patagon ou la preuve des lointains', in
     Céard and Margolin, *Voyager à la Renaissance*, pp. 467–96.
104 Lévi-Strauss, *Tristes Tropiques*, ch. 40, p. 543: 'Every verbal exchange, every
     printed line, establishes communication between people, thus creating an
     evenness of level, where before there was an information gap and conse-
     quently a greater degree of organization.'
105 Thevet, *Cosmographie universelle*, II, fol. 924r.
106 See Olivier Reverdin, *Quatorze Calvinistes chez les Topinambous: Histoire
     d'une mission genevoise au Brésil (1556–1558)* (Geneva: Droz, 1957).
107 Léry, *Histoire d'un voyage* (1580 edn), ch. VI, p. 80.
108 Ibid., p. 82.
109 Ibid., pp. 196–7.
110 Ibid., p. 202. One notices in Thevet, indeed, a confusion of the same order,
     since he applies the title *morbicha* by turns to the 'fearsome' Quoniambec
     and to the old men who formed the council deciding on a war to be un-
     dertaken. Cf. ch. 3, nn. 30–2, above.
111 Ibid.

112 Ibid., p. 196.
113 Ibid., ch. XX, p. 314.
114 Ibid., pp. 283, 285 etc. The *moussacat* is successively defined as the 'goodly head of the family who feeds passing strangers', and as the 'old master of the house'.
115 Ibid. (1580), p. 107.
116 Lévi-Strauss, *Tristes Tropiques*. In the same perspective, one might compare the same author's contribution, 'Eine Idylle bei den Indianern. Über Jean de Léry', in *Mythen der Neuen Welt*, pp. 68–70. As the title suggests, this is less a critical reading of Jean de Léry than a 'retrospective confession' by means of an interposed Léry.
117 See, on this point, the illuminating remarks of Sophie Delpech in her introduction to a new edition of the *Histoire d'un voyage* (Paris: Plasma, 1980), pp. 23–4.

Chapter 5 Cartographics: an Experience of the World and an Experiment on the World

1 On the personality of João Afonso, a pilot of Portuguese origin who was 'bought' by François I and became a naturalized Frenchman, see the decisive study of Luís de Matos, *Les Portugais en France au XVIe siècle. Études et documents* (Coimbra: Acta Universitatis Conimbrigensis, 1952), ch. 1, pp. 1–77. For an evaluation (generally rather severe) of the cosmographical work of Alfonse, see the contribution of Luís de Albuquerque, 'João Afonso (ou Jean Fonteneau) e a sua Cosmografia', in *Les Rapports culturels et littéraires entre le Portugal et la France (Actes du colloque de Paris, 11–16 octobre 1982)* (Paris: Fondation Calouste Gulbenkian, Centre culturel portugais (diff. Jean Touzot), 1983), pp. 101–21.
2 See on this point Numa Broc, *La Géographie de la Renaissance* (Paris: BN–CTHS, 1980), ch. 5 ('Le renouveau de la cosmographie'), pp. 61ff.
3 That is, 5 April 1556, new style. This date figures at the end of the dedicatory epistle addressed to Coligny (see p. 132).
4 Le Testu, *Cosmographie universelle*, fol. 1r. Cf. Thevet, *Cosmographie de Levant* (p. 3, line 23: 'To Monsigneur, Monsigneur François, Conte de La Rochefoucauld'): 'Who is this new Anacharsis or Cosmographer, who after so many Authors, both ancient and modern, can find new things to invent?' (p. 3, line 23).
5 On this legendary character, whom Herodotus invokes in his *Histories* (IV.76), see Hartog, *The Mirror of Herodotus*, pp. 61–84.
6 Le Testu, *Cosmographie universelle*, fol. 1r (see appendix (1) above). Cf. Thevet, *Cosmographie de Levant*, p. 3: 'But I ask them: [p. 4] Was Nature so constrained and subjected by the writings of the ancients, that it was not allowed for her in times to come to give some alternative vicissitude to the things of which they would write?' As is evident, Le Testu's epistle simplifies somewhat the flowery style of Thevet.
7 Michel Mollat du Jourdin, intro. to Michel Mollat and Monique de la Roncière, *Sea Charts of the Early Explorers*, trans. L. le R. Dethan (London: Thames and Hudson, 1984), p. 23: 'This approach, at once descriptive and figurative, recalls the parallel use of *portolano* and *compasso*, portolan chart and rutter, chart and nautical instructions. The problem [which continued

into the simultaneously descriptive and figurative method of the planisphere that accompanied cosmography] retains a dual aspect: calculation and image.'

8   These water-colour copies are today conserved in the Fonds J.-B. d'Anville, BN (Cartes et Plans, Ge DD 2987). For a full list of them, see my catalogue of maps of the *Grand Insulaire*, in Mireille Pastoureau, *Les Atlas français (XVIe–XVIIe siècles)* (Paris: Bibliothèque Nationale, 1984), ch. XXIX, pp. 481–95.

9   Thevet, *Vrais Pourtraits*, II, V, 83 (fol. 482r), 'in which he found himself at great peril and hazard to his life; as, God willing, I hope to show one day in my Grand Insulaire . . .'

10 See ch. 1, nn. 29, 39, above.

11 Thevet's borrowings from the pilotage of the Scottish coasts by Alexander Lindsay, plagiarized and translated by Nicolas de Nicolay are, in the *Grand Insulaire*: 1, fol. 40r ('Isle d'Arren'); fol. 24v ('Isle d'Yla' or Islay). They were based on Nicolay's *Navigation du Roy d'Ecosse Jacques Cinquiesme du nom* (Paris: Gilles Beys, 1583), fols. 4v, 24v. (Translator's note: Nicolay was the French geographer-royal.) Cf. the *Grand Insulaire*, I, fol. 83r ('England and Scotland'). Thevet's 'plagiarization' of the *Voyages avantureux* of Jean Alfonse in several chapters of the *Grand Insulaire* (I, fols. 276v, 280r, 289v–290r) is noted by Roger Hervé, in *Découverte fortuite de l'Australie et de la Nouvelle-Zélande par des navigateurs portugais et espagnols entre 1521 et 1528* (Paris: BN–CTHS, 1982), p. 47, n. 98). But the same commentator notes (fols. 275v and 278v), in the same work, original and 'rather remarkable' descriptions of the 'labyrinth of islands' situated at the opening of the Straits of Magellan on to the Pacific.

   Pigafetta, finally, was used by Thevet – no doubt via the intermediary of Ramusio's Italian translation – especially in the first chapter of Book XXI of Thevet's *Cosmographie universelle*.

12 Pierre Garcie, (known as Ferrande), *Le Grant Routtier, Pillottage, et Encrage de mer* (Poitiers: Jan de Marnef, 1545) (Bib. Mazarine: A 138.45. Réserve), fol. 56r ('Sensuyt de Groye').

13 Jean Alfonse, (João Afonso), *Les Voyages avantureux* (Poitiers: Jan de Marnef, 1559) (BN: Gde Rés. G. 1149-1), fols. 4r, 28r, 32r, 39r, 59r, 64r.

14 Ibid.; copy in private collection, with the antedated ex-libris 'Thevet, 1553' (an anomaly not at all surprising, when one knows the cosmographer's mania for chronology). In this copy, annotated from end to end, the comment 'isle' appears in fols. 4r ('isle de fer'), 9r ('Isle de Calis'), 12r–v, 15r–v ('Isle doleron', 'Isles'), 16r ('baleines / 3. Isles, 'belle / Isle'), 16v ('Isle de Groye', 'Isles', 'Isle de Sein'), 17r ('Isle', 'Isles', 'Isles de bretai/gne'), 18v ('Isle de / Surlingue'), 19r ('Isleaux'), 19v ('Isle / Huich / Isle'), 20v ('Isle', 'lac descosse', 'Islette', 'Isles'), 21r ('I. de Londey'), 21v ('Isle de Main' [of Man], 'Isles', 'Isle'), 22r, 22v, 23r ('30. Isles', 'Isle (s)' repeated three times), 23v ('Isleaux'), 24r ('Isle de Zelande'), 24v (id.), 25v (id.), 26r, 27r ('Isle de Fixlande'), 27v, 28r, 28v ('3. Isles'; 'isles des / exoires'), 29r, 29v ('Isle de St / domingue'; 'Isle de Lucatan' [Yucatan]), 30v ('Perles', 'Isle de la trinite', repeated), 31r ('isles de St / ian et / espagnole'), 31v ('Isle de / môna', 'Isle espagnole'), 32r ('Isle dela / trinite'), 34v ('isles des / moluques'), 35r ('5000. isles' [east of Moluccas]; 'isles que trouvat Magelan', 'nota des isles'), 37r ('isles de / minorque / et maiorque'), 39r ('isles dieres'), 42r ('isle / Corse'), 48v ('isles des / exoires', 'madere'), 49r ('Canaries'), 49v ('isles de / cap deverd'), 57v ('Isles fer'), 59r ('Isle de St / laurans' [Madagascar]), 59v (id.), 60v, 61r, 63r ('Isle de / sel'),

63v ('Isle dieu'), 64r ('isle de / Cochin', '4000 isles' [Maldives]), 64v, 65r, 65v, 66r ('isles de / Clos [of Cloves]'), 67r ('Ydo isle'), 67v.

15   Michiel Coignet, *Instruction nouvelle des poincts plus excellents et necessaires, touchant l'art de naviguer* (Antwerp: Henry Hendrix, 1581) (Bib. du CNAM, Paris: 4e Sa 6), pp. 12, 13. In the chapter 'On Tides' (p. 84), two contradictory annotations ('made' ("fait") and 'not made' "non fait") may give an indication of the progress of Thevet's copying.

The comment 'made' ("fait") is found also in the margin of fol. 51v of the *Voyages avantureux* (private copy, see n. 14), beneath the manuscript headliner, 'Riviere des Rameaux'.

16   Antonio Pigafetta, *Le Voyage et Navigation faict par les Espaignolz es Isles de Mollucques* (Paris: Simon de Colines, c.1526) (Bib. Méjanes, Aix: D. 751), fols. Avi(v), vii, 15r, 17r, 19v, 20v, 36v, 43r, 45r–v, 52r, 53r, 54r–v, 55r–v, 56r, 57v, 61r, 67r–v, 70r. See *The First Voyage Around the World by Magellan, 1518–1521*, ed. Stanley of Alderney (London, 1693; repr. New York: Burt Franklin, 1963).

17   Ibid., fols. Av(v), Avi(v), VIIIr ('folye'), fol. 38, ch. 52 *in fine* ('Thevet ... falsum est'), fols. 54r and 72v ('nichil est'), fol. 43r ('Thevet niquil').

18   Notably, in the copy of Münster's work held in the Bibliothèque du Mans (Histoire F⁰ 54), pp. 1185, 1188: 'ses faulx dit Thevet', 'faulx dit Thevet'. Cf., on p. 13 of the copy of Michiel Coignet's work mentioned in n. 15, the comment 'bourde' (fib or blunder).

19   Jacques-Auguste de Thou, *Historia sui temporis* (Reign of Henri II, 1555), trans. P. du Ryer (Paris, 1659), vol. I, p. 894.

20   Nicolas-Claude Fabri de Peiresc, *Lettres*, ed. Philippe Tamizey de Larroque (Paris: Imprimerie nationale, 1894), vol. 5, p. 304 (lettre XVI, to Lucas Holstenius, dated Aix-en-Provence, 29 January 1629).

21   Abbé A. Anthiaume, 'Un pilote et cartographe havrais au XVIe siècle. Guillaume Le Testu', *Bulletin de géographie historique et descriptive* 1–2 (1911), pp. 11–21.

22   Tommaso Porcacchi da Castiglione, *L'isole piu famose del mondo* (Venice: Simon Galignani and Girolamo Porro, 1572; who also published an augmented edn, with a modified order of description, in 1576).

23   Porcacchi, *L'isole* (1576), pp. 185, 189.

24   Ibid., p. 172.

25   Ibid., p. 193.

26   Ibid., p. 198.

27   On this double destination of Thevet's text, see my study: 'Voyage dédoublé, voyage éclaté. Le morcellement des Terres Neuves dans l'*Histoire* de deux voyages d'André Thevet (c.1586)', *Études françaises* 22, 2 (Autumn 1986), 'Voyages en Nouvelle-France', pp. 17–34, esp. 26–30.

28   Cf. Hilary Louise Turner, 'Christopher Buondelmonti: adventurer, explorer and cartographer', in Monique Pelletier (ed.), *Géographie du monde au Moyen Age et à la Renaissance* (Paris: Éditions du CTHS, 1989), p. 216: 'His aim was, I think, to bring to life for those who could not travel to them those small areas associated with classical mythology and history. Strict geographical accuracy was not his goal, interrelationships were.'

29   *Grand Insulaire*, I, fol. 3v and 4r (nos 2 and 3 of the 'Catalogue des cartes du *Grand Insulaire* d'André Thevet', in Pastoureau, *Les Atlas français*, p. 487).

30   *Grand Insulaire*, p. 25 (cat., no. 1). This engraved frontispiece, its title obscured by an inset, was identified by Marcel Destombes. It is held in

the Cabinet des Estampes, BN, in the Lallemant de Betz collection (Vx I [24] p. 453).

31   See plate I. The portrait is in the BN (Estampes, Rés. 7661; catalogue Robert-Dumesnil no. 495.2). On it, see Marcel Destombes, 'André Thevet, 1504–1592, et sa contribution à la cartographie et à l'océanographie', *Second International Congress on the History of Oceanography* (Edinburgh: Royal Society of Edinburgh, 1972), pp. 123–31.

32   On this divergence of points of view, cf. the discussion of the notion of a 'mediterranean' ocean, similar to an immense Caspian Sea, in ch. 1, above. The cosmographical preliminaries of the atlas, according to Thevet's Preface (fol. 9r), would now be missing at least two maps. Thevet announces there, in fact, 'four figures which serve to show all that is contained in the universe'. Beyond the two surviving 'myparties' of the northern and southern hemispheres, the 'benevolent reader' would on the one hand have had at his disposal a 'figure aiding the understanding of terrestrial and maritime Maps, and of the magnitude of the Universal Earth'; and on the other hand 'the fourth and most difficult figure, that is [to say], that of the rhumbs of winds applied to maritime Charts so that one may voyage all over the Universe, both on land and sea'. The latter map is either lost or was never made, but recalls the *Carta da navigare* concluding the *Isolario* of Tommaso Porcacchi da Castiglione.

33   *Grand Insulaire* (Ms. fr. 15452), I, fols. 142 bis ('Terres Neufves ou Isles des Molues'), 148 bis ('La grand Isle de S. Julien'). On both maps, see my study, 'Terre-Neuve ou la carte éclatée d'après le *Grand Insulaire* d'André Thevet', *Mappemonde*, 87/3, pp. 1–7.

34   See for example the letter Thevet addressed to Ortelius, as 'Geographer and Cosmographer of the Catholic King', around 1586, in *Abrahami Ortelii epistulae*, ed. J. H. Hessels (Cambridge, 1887), vol. 1, pp. 329–30). Thevet writes there, notably, 'As I composed my Cosmography, my Prosopographic History of Famous Men and my Great Isolario, I could not help referring to you and quoting you with honour. In this, I only did the same as you have done, in citing me in your *Theatrum*, Or Synonymy of Geography, and your other works...'

35   I use the term '*bricolage*' in the particular sense given to it by Claude Lévi-Strauss in *The Savage Mind*, ch. 1. See n. 56, below. (Translator's note: I have retained the French word for the same reason, following Lévi-Strauss's translators.)

36   This example is analysed by Danielle Lecoq, in 'La Mappemonde du *Liber Floridus* ou la vision du monde de Lambert de Saint-Omer', *Imago Mundi. The Journal of the International Society for the History of Cartography* 39 (1987), pp. 9–49. In a more general way, symbolism with a didactic value in medieval maps is studied by David Woodward in 'Reality, Symbolism, Time and Space in Medieval World Maps', *Annals of the Association of American Geographers* 75 (4) (1985), pp. 510–21. Cf., by the same author, 'Medieval *Mappaemundi*', in J. B. Harley and D. Woodward (eds), *The History of Cartography*, vol. 1: *Cartography in Prehistoric, Ancient, and Medieval Europe and the Mediterranean* (Chicago: University of Chicago Press, 1987), ch. 18, pp. 330ff.

37   Le Testu, *Cosmographie universelle* (1556), fol. XIXr. See appendix (2), above.

38   Le Testu, *Cosmographie universelle*, fol. XXIr, opposite the map of southern Africa (fol. XXv).

39   Ibid., fol. XXVIIIv, with a map representing the Gulf of Bengal and penin-

sula of Malacca. It is captioned (fol. XXIXr): 'Beyond these, in the furthest part of the Mountains over against the Brazilian people [*jusques a la gent prasi*], are said to be Pygmies who are only one foot high and are infested and molested by cranes. Around the source of the Ganges are found people called *monoceli*, who have only one foot and are able to leap with marvellous levity; also others, called *sciopedes*, who in hot weather lie on their back on the ground and protect themselves from the sun's injuries with the shadow of their foot, which is large enough to cover them. Around the Himalayas (*montagnes Rhiphees*) are found people with dogs' heads, who bark instead of speaking: as Marco Polo the Venetian says.'

40 Ibid., fol. XXXIVv (commentary, fol. XXXVr).

41 Abbé A. Anthiaume, 'Un pilote', p. 45: 'The westward extension of Java-la-Grande seems to favour the hypothesis of a Lusitanian origin of the Southern Lands. The Portuguese, in fact, sought to make these regions fit into the part of the world attributed to them under the Treaty of Tordesillas (7 June 1494).' Cf. R. Hervé, *Découverte fortuite*, p. 27.

42 Le Testu, *Cosmographie universelle*, fol. XXXVr: commentary on the map of fol. XXXIVv showing the 'Islands of Griffins', and, below it, the 'Southern Land', showing a fight between naked and bloody savages. See appendix (3).

43 *Essays*, I, 21. Montaigne, it is true, was here interested only in phenomena that one would today qualify as psychosomatic.

44 Le Testu, *Cosmographie universelle*, fol. XXXVr.

45 Ibid., fol. XXXVIIr: commentary on the map of fol. XXXVIv, 'Mer oceane de l'Inde orientale'.

46 Aujac, Germaine. 'L'Ile de Thulé, de Pythéas à Ptolémée' (communication to the XIIe Congrès International d'Histoire de la Cartographie, Paris, September 1987), in Pelletier, *Géographie du monde*, pp. 181–90.

47 Le Testu, *Cosmographie universelle*, fol. XXXVr (on the 'Southern Land'), transcribed in Appendix (3), above. The same remark is found two pages later (fol. XXXVIIr) apropos the 'Ocean sea of east India': 'And although I have written and noted some names on some capes, it was only to orient the pieces there depicted, and also in order that those who navigate there will be on their guard when in their opinion they are in the vicinity of the said land.'

48 Ibid., fol. XLr.

49 For a definition of this concept, the literal meaning of which is 'that the gaze be able to embrace easily', see the article of Christian Jacob, 'La Mimésis géographique en Grèce antique: regards, parcours, mémoire', in *Sémiotique de l'architecture. Espace et représentation. Penser l'espace* (Paris: Éditions de la Villette, 1982), p. 67.

50 On this notion see Leo Spitzer, *Classical and Christian Ideas of World Harmony. Prolegomena to an Interpretation of the Word 'Stimmung'* (Baltimore, Md.: Johns Hopkins University Press, 1963).

51 On the circumstances of this cartographical fiction and its durability, see, apart from Anthiaume, 'Un pilote', pp. 45–6, R. Hervé's *Découverte fortuite*, pp. 27ff, and David Fausett, *Writing the New World: Imaginary Voyages and Utopias of the Great Southern Land* (Syracuse, N.Y.: Syracuse University Press, 1993), ch. 1.

52 BN: Cartes et Plans (Rés. Ge AA 625). This world map in two hemispheres (one leaf on felt, MS in blackish brown, 1180 × 790 mm) bears the following inscription: 'This Map was portrayed in all perfection, both of latitude and longitude, by me, Guillaume Le Testu, Pilot-Royal, a native of the French

city of grace [Le Havre]. In favour of that noble and illustrious person, Pierre de [Coutes], Sieur de la [Chapelle] du Pré et du Bouschet, Cap[tai]n Ordinary of the King in his Western Navy; and was completed on the 23rd day of May 1566.' For a commentary on this map (formerly in the archives of the Ministry of Foreign Affairs, now in the Bibliothèque Nationale), see A. Anthiaume, 'Un pilote', pp. 59–70.

53 Le Testu, map cited (n. 52), left-hand side: 'Because this coast and land have until now not been duly discovered, I prefer to leave it imperfect rather than to add to this true map any *mensonge*.' Cf. (ibid.), apropos of Labrador, 'I have left this coast imperfect, although some complete it, until it is more amply discovered.'

54 Le Testu, map cited (n. 52), cartouche over the North Pacific (previously noted by Anthiaume): 'In the Year 1527, by command of the Emperor Charles of Austria, three ships were sent out from the port of Ciuattancio to discover the Moluccas, whose captain was Don Alvaro Savedra; who found on his expedition that there were 1,400 leagues from the above-mentioned port to the Moluccas, where he and his crew remained on account of contrary weather, with the exception of eight men who were taken prisoner at Malacca by the Portuguese, who have told of this.'

55 Le Testu, map cited (n. 52), inscription in a cartouche at bottom left, not far from the 'Strait of Magellan' (Détroit de Magellan).

56 Lévi-Strauss, *The Savage Mind*, p. 24: 'I have so far only considered matters of scale which . . . imply a dialectical relation between size (i.e. quantity) and quality. But miniatures have a further feature. They are "man made" and, what is more, made by hand. They are therefore not just projections or passive homologies of the object: they constitute a real experiment with it.'

57 Le Testu, *Cosmographie universelle*, fol. LVIIv and commentary, fol. LVIIIr (see appendix (9), above). This map (water-colour on paper, 53 × 37 cm) is reproduced in La Roncière and Mollat, *Les Portulans* (Fribourg: Office du Livre; Paris: Nathan, 1984), plate 50.

58 Le Testu, *Cosmographie universelle*, fol. LVIv and commentary, fol. LVIIr, transcribed in appendix (8), above. This 'Terra Nova' is an archipelago in three main parts, with a diagonal line strongly marked by the *monts des Granches*. It is plate 49 in La Roncière and Mollat, *Les Portulans*. The incessant, complex metamorphoses in the cartography of Newfoundland in the course of the sixteenth century are minutely traced by Henry Harrisse, in *Découverte et évolution cartographique de Terre-Neuve*; his ch. 16 ('Nouvelles cartes dieppoises', pp. 260ff) analyses the contribution of Le Testu and Thevet to this evolution.

59 An expression of this 'deception' is found in Thevet, when he opposes, in a eulogy of navigation, 'inconstant and uncertain islands' to 'firm, good and fertile land' (*Histoire de deux voyages*: 'The author's embarkation', ch. 1, fol. 1v).

60 This definition, apparently contradictory, of the 'round' form of the rondo (which, as we know, is characterized by the return in a refrain of the first hemistich or entry) is borrowed from Pierre Le Fevre (Fabri), *Le Grand et Vrai Art de pleine rhétorique* (Rouen, 1521; repr., with notes and glossary, by A. Héron, Rouen: A. Lestringant, 1890). See, on this subject, the 'note de poétique' in my edition of Clément Marot, *L'Adolescence clémentine* (Paris: Gallimard (Poésie), 1987), pp. 335–42.

61 The 'Ille des grandz hommes' that Le Testu situates north of the Straits of Magellan (*Cosmographie*, fol. XLIXv) may correspond to the Malvinas or

Falkland Islands. According to Roger Hervé (*Découverte fortuite*, pp. 25, 68), the 'île de la Joncade' in the South Pacific would be New Zealand.

62 On La Popelinière's geopolitical myth of a Southern Land, see the study of Erich Hassinger, 'Die Rezeption der Neuen Welt durch den französischen Späthumanismus (1550–1620)', in Wolfgang Reinhard (ed.), *Humanismus und Neue Welt*, pp. 111–14, 128–32. See also Hassinger's *Empirisch-rationaler Historismus. Seine Ausbildung in der Literatur Westeuropas von Guicciardini bis Saint-Evremond* (Berne and Munich: Francke, 1978), pp. 26–35.

63 Thus, the *Theatrum orbis terrarum* of Abraham Ortelius (Antwerp: Christopher Plantin, 1570) provided Thevet with models for the islands of Islay, Mull, North and South Uist, and Lewis in the Hebrides ('Scotland' map, no. 10), and for the 'Britannic Isles' (map no. 9) in the vignette of the *Grand Insulaire* representing 'England and Scotland' (cat. no. 40, fol. 82 ter), again oriented with north facing left. The same might be said of 'Hibernia or Ireland' (Ortelius, 13; cf. *Grand Insulaire*, 42); of the Islands of Voorn, Walcheren and Zuid Beveland, which passed from the map of 'Zeeland' (Ortelius, 39) into Thevet's archipelago in the *Grand Insulaire* (7, 9, 10 respectively); and so on. As for Mercator, his famous world map of 1569 was used by Thevet in the four continental maps figuring at the beginning of each of the volumes of the *Cosmographie universelle* of 1575.

64 Thevet, *Grand Insulaire*, fol. 142 bis. See plate IX.

65 Jacques Cartier, *The Voyages of Jacques Cartier*, trans. and ed. H. P. Biggar (Ottawa: Publications of the Public Archives of Canada, 1924), p. 9: 'and our men found one as big as a calf and as white as a swan that sprang into the sea in front of them.' (Translator's note: John Florio's 1580 translation, *A Shorte and Briefe Narration of the Two Navigations to Newe Fraunce* (London: Bynneman), is reprinted by Walter J. Johnson, 1975.)

66 Thevet, *Grand Insulaire*, fol. 145r: 'As for the island that you see in my maps surnamed with my name, Thevet, this comes about from the fact that I was the first among my companions to set foot on it when we put in there in a skiff to get some fresh water ...' The same Canadian island is mentioned in passing in his *Cosmographie universelle* (II, fol. 1009v), but without explanation of how it came to be named.

67 *Grand Insulaire*, fol. 260r. The engraved map constitutes fol. 259 bis. The Brazilian Island of Thevet was earlier evoked in his *Cosmographie universelle* (II, fol. 1015r): 'I shall not forget, here, an island that is situated eight degrees and ten minutes beyond the equator in a south-south-easterly direction, which had not previously been discovered by anyone nor marked on any map; at which I landed with some others in a skiff to find fresh water. After the others had anchored – for we had a contrary wind – we made a pact that the first to reach the island would have it named after him. Which happened to be me; so that it was named Thevet Island. The sole inhabitants and governors of it were birds of various plumages and sizes ...'
    On these two 'Thevet Islands', see my study: 'Nouvelle-France et fiction cosmographique dans l'oeuvre d'André Thevet', *Études littéraires* 10, nos. 1–2 (April–August 1977), pp. 145–73.

68 Thevet, *Cosmographie universelle*, II, Book XXI, ch. 1, fol. 907 (verso of fol. 906). See ch. 1, above.

69 *Grand Insulaire*, I, fol. 245r ('Du Goulphre ou Riviere de Guanabara'). The map (14.6 × 18.3 cm) corresponding to this chapter is in the J.-B. Bourguignon d'Anville Collection, BN (Cartes et Plans, Rés. Ge DD 2987, no. 9480).

70 Jacques de Vaudeclaye, *Le vrai pourttraict de Geneure et du Cap de Frie*, MS

in colour on vellum (645 × 485 mm), in BN (Cartes et Plans, Rés. Ge C 5007). This document is reproduced and discussed in Christian Jacob and Frank Lestringant, *Arts et légendes d'espaces* (Paris: Presses de l'École Normale Supérieure, 1981) pp. 243ff; also in La Roncière and Mollat, *Les Portulans*, plate 61 and pp. 240–1.

71 On this polemic, see my article 'Fictions de l'espace brésilien à la Renaissance: l'exemple de Guanabara', in Jacob and Lestringant, *Arts et légendes d'espaces*, pp. 205–56.

72 The expression is from Gilles Lapouge, *Equinoxiales* (Paris: Le Livre de Poche, 1977), p. 55.

73 *Grand Insulaire*, II (Ms. fr. 15453), fols. 102 bis ('Escueil ou Isle de Strongile a p[rese]nt Peogola') and 103v (commentary).

74 Ibid. II, fol. 93 bis ('L'Isle de Church of Curco)'). The commentary is on fol. 94v.

75 Ibid. II, fols. 56 bis ('Le Caloiero de Nisaro dit Panegea') and 90 bis ('Caloiero d'Andros / dit le bon vieillart'). On this doublet, see my study 'Fortunes de la singularité à la Renaissance: le genre de l'Isolario', *Studi francesi* 84 (September–December 1984), pp. 415–36, esp. 430–5.

76 *Grand Insulaire* (Ms. fr. 15452), I, fols. 145v ('Isle de Roberval'), 153r ('Isle des Démons'), 404r ('Isle de la Damoiselle'). On the legend of Marguerite de Roberval and its literary fortunes, see Arthur P. Stabler, *The Legend of Marguerite de Roberval* (Pullman: Washington State University Press, 1972).

77 *Grand Insulaire*, I, fol. 147r–v. Cf. Schlesinger and Stabler, *André Thevet's North America*, pp. 235–6. The attacks on Rabelais were suppressed in Thevet's later *Description de plusieurs isles* (manuscript, c.1588: BN: Ms fr. 17174, fol. 95r–v).

78 *Cosmographie universelle*, I, fols. 443v–446r. Cf. ch. 4, above, nn. 47–57.

79 *Grand Insulaire*, I, fol. 268r. Roger Hervé (*Découverte fortuite*, p. 124) identifies these enigmatic islands with the Falkland archipelago.

80 On this encounter, see ch. 4, n. 103, above.

81 *Grand Insulaire*, I, fol. 275 bis.

82 Ibid., fol. 277v. The map of this island is missing from the manuscript.

83 Ibid., fol. 278v.

84 Ibid., fol. 280v. The map is missing.

85 Hervé (*Découverte fortuite*, pp. 42–3; trans. as *The Chance Discovery of Australia* ... by John Dunmore Palmerston North: Dunmore Press, 1987) discusses the itinerary he believes was followed by the caravel *San Lesmes*, separated from the fleet of Garcia Jofre de Loaysa in February 1526 while rounding South America, and 'for which the only fairly definite source is ... Alfonse's first ruttier ... I consider that there can be only one ship mentioned twice in almost the same terms and in only two passages of the *Voyages avantureux*. The Norman cartographers, Alfonse's contemporaries, do not seem to have been sure which voyage this was, but they knew of sundry navigations in the waters of the "Magellanic Sea" and their comments refer to them discreetly. Later, the cartographer André Thevet will be bold enough to reproduce in full the relevant passage of the *Voyages avantureux*, with his own embellishments. He suspected that a voyage aiming "to make straight for the Moluccan Islands" may have taken place "twelve years since, when a certain ship ploughed that frightful sea". The old geographer had solid reasons for his comments because he had had occasion to personally consult various seamen who had taken part in explorations of the "Magellanic waters", as he tells us on several occasions. Aside from textual

borrowings from Alfonse and in spite of tedious, repetitive passages, such an impression of truth and visual freshness emerges from his descriptions ... that one is left wondering which seamen he had consulted out of those who had tried to cross [Magellan's] straits between 1520 and 1540 – Magellan, Loaysa, Alcazova or Camargo.'

Thevet evokes several times 'the memories [or memoirs: *mémoires*] I have had of several old Portuguese, Spanish and other Pilots' (*Grand Insulaire*, I, fol. 227v). Again, apropos of the 'Isles Geantées' of Sanson, off Patagonia, he mentions 'an old Portuguese Captain and good Pilot, whom I found in the city of Lisbon' (ibid., fol. 269v). But he avoids spelling out his debt to the deceased Alfonse.

(Translator's note: the Hervé passage above is quoted at greater length than in the original, partly to convey fully its sense, and partly because it seems to imply the possibility that Thevet's source may not, in fact, have been Alfonse alone. This problem is discussed in my *Writing the New World*.)

86   *Grand Insulaire*, I, fol. 151r; and *Description de plusieurs isles*, fol. 97r.
87   *Grand Insulaire*, I, fol. 149v (cf. *Description de plusieurs isles*, fol. 100v): 'For that, however, one should not tread underfoot (neglect, and tread underfoot) the hope of its fertility, given that Holland, Zealand and other regions, although lying in swamps, have not delayed in exploiting their fecundity.' The transcription of this passage from the *Grand Insulaire* by Schlesinger and Stabler (*André Thevet's North America*, pp. 240–1), should be used with caution.
88   I borrow this terminology from Jack Goody, *La Raison graphique. La domestication de la pensée sauvage* (Paris: Éditions de Minuit, 1979), pp. 149, 164–9. (Translator's note: these terms, *événementielle* and *récapitulative*, are Goody's translator's. Goody himself, in *The Logic of Writing and the Organization of Society* (Cambridge: Cambridge University Press, 1987, pp. 162, 181–5), speaks of historical 'instances' and their accumulation and comparison through writing-processes; and of 'redistributive' (or 'primitive') economic systems.)
89   Thevet, *Histoire de deux voyages*, fols. 149r–155r: 'De Canada, et terre-neufve'; 'La route et dangers de la terre de Canada'; 'S'ensuit un petit Dictionnaire de la langue des Canadiens.'
90   *Grand Insulaire*, I, fol. 276v.
91   Cf. ch. 1, above.
92   *Grand Insulaire*, I, Preface (fol. 6r).
93   La Popelinière, the 'singer of an Antarctic epic' (*chantre de l'Antarctide*) of whom Numa Broc writes, was inspired by the recent voyages of Drake as well as by the reports of a Portuguese pilot, Bartolomeu Velho, and by a cosmographer of Italian origin, André d'Albaigne. See, on this point, Broc's synthesis: 'De l'Antichtone à l'Antarctique', in *Cartes et figures de la terre* (Paris: Centre Georges Pompidou, 1980), pp. 136–49; and the still indispensable volume of Armand Rainaud, *Le Continent austral. Hypothèses et découvertes* (Paris: Armand Colin, 1893).

Epilogue: The End of Cosmography

1   Thevet, *Cosmographie universelle*, II, Book XXI, ch. 1 (fol. 903v): 'I have obtained for myself two of their coats, made from the skins of animals, in a colour of which I have not seen the like in that land or any other I have

visited. When I put one on, more than a metre of it hung down on the ground . . .' Again, 'They had fired at the ships volleys of arrows, that were so well fixed that they could only with difficulty be pulled out of the sides of the ships, so far in had they penetrated; and instead of iron, they were fixed and bound with animal bones and very sharp stones; some of which I recovered and have in my cabinet in Paris . . .' Cf. also fol. 905r: 'I recovered a bow and some arrows which came from their land, and which I presented to the late and lamented Anthoine, King of Navarre.'

All these objects had the function of confirming the veracity of the tale of a voyage to a distant land. They had the value of media(tors), transferring 'over to this side' the admirable and radiant presence of the 'beyond'. See, on the subject of such intermediary objects, my study 'La flèche du Patagon ou la preuve des lointains: sur un chapitre d'André Thevet', in Céard and Margolin, *Voyager à la Renaissance*, pp. 467–96.

2 If the *Grand Insulaire* was partly recopied and put in order in the *Description de plusieurs isles*, the *Histoire de deux voyages* represents the final state of a relation initially called *Second voyage*, the core of which is constituted by Book XXI of the *Cosmographie universelle*. Since the *Description de plusieurs isles* and the *Second Voyage* are of the same writing, it appears that the *Histoire de deux voyages* was the last of the projects Thevet worked on.

3 Rabelais, *Third Book*, chs 49–51; cf. Pliny, *Natural History* XIX.1.

4 In the novella 'Pierre Ménard, Author of Don Quixote', in his *Ficciones*, Borges creates a symbolist poet from Nîmes who sets out, not to copy – that would be too easy – but to rewrite (apart from the signature) *Don Quixote*. This 'stupefying' effort of plagiarism leads to a statement of the principle of erroneous attributions at the close of this philosophical story: 'Would not the attributing of *The Imitation of Jesus Christ* to Louis-Ferdinand Céline or James Joyce be a sufficient renovation of the tenuous spiritual counsels of the work?' (Jorge Luis Borges, *Fictions*, trans. Anthony Bonner, ed. and intro. Anthony Kerrigan, London: John Calder, 1985, pp. 42–51).

5 See André Tournon, *Montaigne: la glose et l'essai* (Lyon: Presses Universitaires de Lyon, 1983).

6 An idea borrowed from Jean Céard, in 'Les transformations du genre du commentaire', *Actes du colloque l'Automne de la Renaissance (1580–1630)* (Paris: Vrin, 1981), pp. 101–15.

7 Thevet, *Le Livre contenant la description de tout ce qui est comprins soubz le nom de Gaule*, BN: Ms. fr. 4941.

8 Ibid., fol. 5r, interlinear annotation: 'as I am led to believe'; fol. 4r (on Toulouse): 'Which I find difficult to believe, since I have not seen any signs of such great antiquity'.

9 François Rigolot, *Le Texte de la Renaissance. Des Rhétoriqueurs à Montaigne* (Geneva: Droz, 1982).

10 Around 1872, according to his *Carnets de travail*, Gustave Flaubert consulted Thevet's *Cosmographie universelle* while preparing the final version of the *Temptation of St Anthony*, published in April 1874. Thevet's teratological iconography was at the origin of the 'birds that feed on wind', the theory of which forms the story's penultimate movement, before the 'sea-creatures' and the final materialist crisis. See Flaubert, *Oeuvres complètes* (Paris: Club de l'Honnête Homme, 1973), vol. 8, pp. 383–4 (*Carnet 16*); or *Carnets de travail*, ed. Pierre-Marc de Biasi (Paris: Balland, 1988), p. 668.

11 Cf. Claude-Gilbert Dubois, *L'Imaginaire de la Renaissance* (Paris: Presses

Universitaires de France, 1985), pp. 83ff, and Leo Spitzer, *Classical and Christian Ideas of World Harmony* (Baltimore, Md.: Johns Hopkins University Press, 1963).

12 Samuel Purchas, *Hakluytus Posthumus or Purchas His Pilgrimes* (London, 1625; reissued Glasgow: James MacLehose, 1906), vol. 15, pp. 412ff: 'The History of the Mexican Nation, described in pictures by the Mexican Author explained in the Mexican language; which exposition translated into Spanish, and thence into English, together with the said Picture-historie, are here presented.' Purchas had, in fact, tried to reproduce *in extenso* the iconographic part of the Aztec *Codex.*

13 Alfred Métraux, *La Religion des Tupinamba et ses rapports avec celle des autres tribus Tupi-Guarani* (Paris: E. Leroux, 1928), pp. 239–52, where are transcribed fols. 53–62v of Thevet's *Histoire de deux voyages*; Lévi-Strauss, *Tristes Tropiques*, p. 456; Clastres, *Le Grand Parler*, pp. 95–9; Christian Duverger, *L'Origine des Aztèques* (Paris: Seuil, 1983), passim, esp. pp. 35–6, 109–10. See in addition the bibliography of modern (post-1800) editions of Thevet given below.

14 Lescarbot, *Histoire de la Nouvelle France*, Book 2, ch. 29, pp. 425–8: 'Censure of certain writers who have written about New France'. Belleforest was also taken to task; but Thevet was spared, and elsewhere (pp. 208–9) praised for his prospective fiction of Henryville.

15 Jean Alfonse (de Saintonge), *Les Voyages avantureux* (Poitiers: Jan de Marnef, 1559), fol. 62v (the tomb of Mahomet suspended in the air by a lodestone); 67v (the gold-seeking ants, or Alibifors); 27v: (apropos of Newfoundland) '. . . then steer west for more than a hundred leagues. All the people of this land have a tail.'

Appendix: Extracts from Guillaume La Testu's *Cosmographie Universelle*

1 Felines such as the spotted lynx, snow leopard or mountain panther (trans.).
2 Randle Cotgrave's *Dictionnarie of the French and English Tongues* (London, 1611; ed. and intro. W. Woods, Columbia: University of South Carolina Press, 1950) gives for *loup-cervier* 'A kind of white wolfe, or beast ingendred between a Hind and a wolfe, whose Skinne is much esteemed by great men, yet some . . . imagine it rather to bee the spotted Linx, or Ounce.' Cf. n. 1, above (trans.).

Bibliography of Works by André Thevet

Early Editions and Manuscripts

For a detailed bibliography of Thevet's works and the location of existing copies, see my critical edition of the *Cosmographie de Levant* (Geneva: Droz, 1985), pp. LXXV–LXXVIII, XCIII–XCVI); as well as my work *André Thevet, cosmographe des derniers Valois* (Geneva: Droz, 1991).

Cosmographie de Levant, par F. André Thevet d'Angoulesme. Lyon: Jean de Tourmes and Guillaume Gazeau, 1554. Small in–4°, 214 pp. + 8 fols., 25 woodcut figures and portrait of author as Franciscan monk.

Cosmographie de Levant, par F. André Thevet d'Angoulesme. Revue et augmentee de plusieurs figures. Lyon: Jean de Tourmes and Guillaume Gazeau, 1556. Small in–4°, 218 pp. + 7 fols., 34 figures and portrait of author.

Cosmographie de Levant, par F. André Thevet, d'Angoulesme. Antwerp: Jean Richart, 1556. In–8°, 158 fols., figure.

Cosmographia Orientis. Das ist Beschreibung desz gantzen Morgenlandes . . . in Teutsche Sprache versetzt und mit nützlichen marginalibus *vermehret durch Gregor Horst.* Giessen: Caspar Chemlin, 1617. In–4° of 4 unnumbered fols. + 216 pp., frontispiece and figure in text; 12 extra-text plates.

Les Singularitez de la France Antarctique, Autrement nommée Amerique: et de plusieurs Terres et Isles decouvertes de nostre temps. Paris: The Heirs of Maurice de la Porte, 1557 and 1558. In–4° of VIII + 166 fols. + table; 41 woodcut engravings.

Les Singularitez de la France Antarctique . . . Antwerp: Christopher Plantin, 1558. In–8° of VIII + 164 fols.; 41 woodcut engravings (reduced copies of those in preceding edn).

Historia dell' India America, detta altramente Francia Antartica . . . tradotta di Francese in Lingua Italiana, da M. Giuseppe Horologgi. Venice: Gabriel Giolito De' Ferrari, 1561. In–8° of 16 fols. + 363 pp. (reissued 1583, 1584).

The New found worlde, or Antarctike, wherein is contained wonderful and strange things, as well of humaine creatures, as Beastes, Fishes, Foules, and Serpents, Trees, Plants, Mines of Golde and Silver: garnished with many learned aucthorities, travailed and written in the French tong, by that excellent learned man, master Andrewe Thevet. And now newly translated into Englishe, wherein is reformed the errours of the auncient Cosmographers. London: Henry Bynnemann, for Thomas Hacket, 1568. In–8° of 8 unnumbered fols. + 138 fols. + 2 fols. of contents (British Library: 798.c.34 and G. 7107); includes Hacket's dedicatory epistle to Sir Henrie Sidney.

La Cosmographie universelle d'André Thevet cosmographe du Roy. Illustree de

*diverses figures des choses plus remarquables veuës par l'Auteur, et incogneuës
de noz Anciens et Modernes.* Paris: Pierre l'Huillier and Guillaume Chaudière,
1584, 4 vols (2 + 2) in-fol., 1 + 1,025 numbered fols., tables; 228 woodcuts incl.
the frontispiece repeated seven times; and 4 maps of the continents.

*Les Vrais Pourtraits et Vies des Hommes illustres Grecz, Latins et Payens, recueillez
de leurs tableaux, livres, medalles antiques et modernes. Par André Thevet
Angoumoysin, Premier Cosmographe du Roy.* Paris: Jacques Kerver's Widow
and Guillaume Chaudière, 1584, 2 vols bound together in-fol., of 16 numbered
fols. + 664 numbered fols. + 18 fols. of Contents; 224 copper-plate portraits.

*Histoire des plus illustres et scavans hommes de leurs siecles. Tant de l'Europe, que
de l'Asie, Afrique et Amerique. Avec leurs portraits en taille-douce, tirez sur les
veritables originaux, par A. Thevet, historiographe. Divisé en huit tomes.* Paris:
François Mauger, 1671, 9 vols. in–12° (8 of text, 1 of portraits).

*Prosopographia: Or, some select Pourtraitures and Lives of Ancient and Modern
Illustrious Personages. Collected out of their Pictures, Books, and Medals.
Originally compiled and written in French by Andrew Thevet, Chief
Cosmographer to Henry the third, King of France and Poland. Newly trans-
lated into English by some learned and eminent Persons; and generally by Geo.
Gerbier, alias D'Ouvilly, Esq.* London: Abraham Miller, for William Lee, 1657.
In-fol. of 1 unnumbered fol. + 76 pp. in 20 chapters; 19 engraved portraits.
This composition forms an addendum to Plutarch's *Lives*, translated (from
the French version by Jacques Amyot) by Thomas North.

*Prosopographia: Or, some select Pourtraitures and Lives of Ancient and Modern
Illustrious Personages ... And now also in this Edition are further added the
Pourtraitures and lives of five other selected eminent Persons, of Ancient and
Modern Times; newly translated into English, out of the Works of the said
famous Andrew Thevet, by a learned and eminent Person.* Cambridge: John
Hayes, for William Lee in London, 1676. One vol. in-fol. of 2 unnumbered
fols. + 91 pp., as an addendum to Sir Thomas North's Plutarch. The comple-
mentary series of five portraits and lives (pp. 73–91), including Ferdinand
Cortez, Basil the Duke of Muscovy, Sebastian I of Portugal, Quoniambec and
Columbus, is introduced by its own title-page (p. 73, with blank verso).

*Le Grand Insulaire et Pilotage d'André Thevet Angoumoisin, Cosmographe du
Roy. Dands lequel sont contenus plusieurs plants d'usles habitées, et deshabitées,
et description d'icelles,* MS (1586–7). Bibliothèque Nationale: Ms. fr. 15452–
15453 (fonds Séguier-Coislin; Saint-Germain français 654). 2 MS vols of 413
and 230 fols. (358 × 225 mm); 84 copper-plate maps inserted at corresponding
chapter headings. For a description of the cartographical corpus linked to the
Grand Insulaire, see my 'Catalogue des cartes du *Grand Insulaire* d'André
Thevet', in Mireille Pastoureau, *Les Atlas français (XVIe–XVIIe siècle),* Paris:
Bibliothèque Nationale: 1984, pp. 481–95.

Description de plusieurs Isles, par M. André Thevet (1588). BN: Ms. fr. 17174
(fonds Séguier-Coislin: Saint-Germain français 655). One vol. of 145 fols. (345
× 220 mm). This is a partial ordering of the *Grand Insulaire.* The 51 chapters
deal with islands in the North Sea, the English Channel and the Atlantic.

*Histoire d'André Thevet Angoumoisin, Cosmographe du Roy, de deux voyages
par luy faits aux Indes Australes, et Occidentales. Contenant la façon de vivre
des peuples Barbares, et observation des principaux points que doivent tenir en
leur route les Pilotes, et mariniers, pour eviter le naufrage, et autres dangers de
ce grand Ocean, Avec une response aux libelles d'injures, publiées contre le
chevalier de Villegagnon* (1587–8). BN: Ms. fr. 15454 (collection Séguier-Coislin;

Saint-Germain français 656). One MS vol. of 167 fols. (355 × 220 mm). The writing is the same as that of the *Description de plusieurs Isles.*
The 'Response aux libelles d'injures publiées contre le chevalier de Villegagnon' inserted into fols. 108v–110v is a faithful copy (apart from two details) of an apologetic booklet published by Villegagnon with André Wechel in 1561 (BN: 8° Lb 33. 388). The *Histoire de deux voyages* was a riposte to Jean de Léry's *Histoire d'un voyage* (Geneva: Antoine Chuppin, 1578, 1580, 1585), at the same time as being an amplification of the cosmographer's Brazilian account (*Singularitez de la France Antarctique* and, especially, *Cosmographie universelle*, Book XXI).
Second Voyage d'André Thevet, dans les Terres Australes et Occidentales (1587?). BN: Ms. fr. 17175 (collection Séguier-Coislin; Saint-Germain français 657). One vol. of 178 fols. (300 × 200 mm). Thevet's text begins only at fol. 11r. This was a contemporary copy of the *Histoire de deux voyages*, with handwritten corrections by the author.

Modern Editions (after 1800)

1858

1 *Cosmographie moscovite, par André Thevet.* Collected and published by Prince Augustin Galitzin. Paris: J. Techener, 1858. One vol. in–16°, xvi–181 pp. This is an uncriticized reissue of 5 chapters of the *Cosmographie universelle* (II, XIX, 8–12) and one from the *Vrais Pourtraits et Vies des Hommes illustres* (II, V, 56).

1878

2 *Les Singularitez de la France Antarctique.* New edn with notes and commentary by Paul Gaffarel. Paris: Maisonneuve, 1878. One vol. in–8° of lxiv + 459 pp.; vignettes based on engravings in the original edition.

1881

3 *La Grande et Excellente Cité de Paris*, intro. and notes by Abbé Valentin Dufour. Paris: A. Quantin (Collection des anciennes descriptions de Paris, no. 5), 1881. One vol. in–8° of xxiv + 55 pp.; figures and a portrait, incl. the Emmaüs tapestry in the Gaignières collection (BN: Cabinet des estampes). Reissue of a chapter of the *Cosmographie universelle* (II, XV, 5).

1882

4 'The Cosmography of the Fraudulent Thevet', *Magazine of American History, with Notes and Queries*, 8, part 1 (January 1882), pp. 130–8. A publication of the pages from the *Cosmographie universelle* (II, XXIII, 3) concerning the future New England, or Norambègue.
5 *Le Grand Insulaire et Pilotage d'André Thevet (Isle de Haity ou Espaignole; Isle Beata; Isles du Chef de la Captive).* In the appendix (pp. 153–81) of the *Discours de la navigation de Jean et Raoul Parmentier de Dieppe*, ed. Charles

Schefer. Paris: Ernest Leroux (Voyages et documents pour servir à l'histoire de la géographie), 1883. Extracts from the *Grand Insulaire* (I, fol. 183 ter–194r).

1890

6 *Le Grand Insulaire et Pilotage d'André Thevet (12 chs: Venice, Zarre, Bua, Corfou, Duché, Grande Céphalonie, Petite Céphalonie, Sapience, Cérigo, Scarpante, Nissare, Cypre)*, in appendix (pp. 245–309) of the *Voyage de la Terre Sainte composé par Maistre Denis Possot et achevé par Messire Charles Philippe*, ed. Charles Schefer. Paris: Ernest Leroux (Voyages et documents pour servir à l'histoire de la géographie), 1890. Extracts from the *Grand Insulaire* (I, passim).

7 *Jeanne d'Arc, par André Thevet. Extrait de ses 'Vrais Pourtraits et Vies des Hommes illustres' (1584), avec une note sur les armes de la Pucelle*, ed. Pierre Lanéry d'Arc. Orléans: H. Herluison, 1890, 43 pp. Extracts from the *Vrais Pourtraits* (II, IV, 25).

1903

8 'De l'isle de Madagascar, autrement de Saint-Laurent, par A. Thevet (1558)'; 'De l'isle d'Albargra ou Madagascar et du deluge advenu en icelle, par A. Thevet (1575)'; 'Des habitans de Madagascar ou de S. Laurent', in *Collection des ouvrages anciens concernant Madagascar*, ed. Alfred and Guillaume Grandidier. Paris: Comité de Madagascar, 1903, vol. 1, pp. 105–9, 118–26, 127–34, 148–55. Extracts from the *Singularitez de la France Antarctique*, ch. 23; *Cosmographie universelle* (I, IV, chs 4 and 5); *Grand Insulaire* (I, fols. 342v–344v). The editor notes (p. 51, n. 1) that a whole part of the *Grand Insulaire* description is borrowed from F. de Belleforest's *Cosmographie universelle* (II, XXVI, col. 2011).

1905

9 'Isle d'Alopetie dicte des renards'; 'Ce premier dictionnaire en langue moscovite appartient à M. André Thevet Premier Cosmographe du Roy', appended (pp. 19–33) to Paul Boyer, 'Un vocabulaire français–russe de la fin du XVIe siècle, extrait du *Grand Insulaire* d'André Thevet, manuscrit de la Bibliothèque Nationale, publié et annoté par P.B.', an extract from *Mémoires orientaux. Congrès de 1905 (Alger)*. Paris, Ernest Leroux, 1905, pp. 9–18. An extract from the *Grand Insulaire* (II, fols. 211r–224r).

10 '*Histoyre du Mechique*. Manuscrit français inédit du XVIe siècle', ed. Édouard de Jonghe, *Journal de la Société des américanistes de Paris*, n.s. II, no. 1 (1905), pp. 1–43. A publication of the Bibliothèque Nationale's Ms. fr. 19031, fols. 79–88.

1910

11 'Thevet's Description of Chios', the Appendix to F. W. Hasluck, 'The Latin Monuments of Chios', *Annual of the British School at Athens*, XVI (1909–10 session), pp. 137–84. An extract from the *Grand Insulaire* (II, fols. 162v–163v).

1926

12   *Jean Guttemberg, Inventor of Printing. A Translation by Douglas C. McMurtrie of the Essay in A. Thevet's* Vies des hommes illustres. *Paris, 1589 [sic].* New York, 1926. In–4°, 8 pp. A translation of the *Vrais Pourtraits* (II, VI, ch. 97).

1928

13   'Mythes des Tupinamba recueillis par Thevet de la bouche du roi Quoniambec et d'autres vieillards lors de ses voyages à Rio de Janeiro en 1550 et 1555'; 'L'anthropophagie rituelle des Tupinamba', appendices (pp. 225–39 and 239–52) to Alfred Métraux, *La Religion des Tupinamba et ses rapports avec celle des autres tribus Tupi-Guarani.* Paris: Ernest Leroux, 1928. Extracts from the *Cosmographie universelle* (II, fols. 913–920v) and from the *Histoire de deux voyages* (fols. 53–62v).
14   'Les Indiens Waitaka (à propos d'un manuscrit inédit du cosmographe André Thevet)', ed. Alfred Métraux, *Journal de la Société des Américanistes de Paris*, n.s. XXI, fasc. 1 (1929), pp. 107–26. A publication (pp. 121–4 of this article) of fols. 101 and 114–16 of the *Histoire de deux voyages*.

1933

15   'Des montagnes qui sont en la contrée de Queurevrijou, païs des Tapoüys, joignant la riviere de Potijou'; 'Des contrées de Ouyana, Achyrou, et des singularités d'icelle'; 'Des provinces de Tararijou et Daritama en la province de Margana, et gouverneurs d'icelles', pp. 33–40 of Alfred Métraux, 'Un chapitre inédit du cosmographe André Thevet sur la géographie et l'ethnographie du Brésil', *Journal de la Société des Américanistes de Paris*, n.s. XXV, fasc. 1 (1933), pp. 31–40. A publication of fols. 33–7 of the *Histoire de deux voyages*.

1944

16   *Singularidades da França Antártica a que outros chaman de America. Prefacio, tradução e notas do Prof. Estevão Pinto.* São Paulo: Cia Editora Nacional, 1944. In–8°, 502 pp.

1953

17   *Les Français en Amérique pendant la deuxième moitié du XVIe siècle, I: Le Brésil et les Brésiliens par André Thevet*, selection with notes by Suzanne Lussagnet, intro. Charles-André Julien. Paris: Presses Universitaires de France (Les Classiques de la colonisation), 1953. viii + 346 pp. Extracts from the *Cosmographie universelle*, vol. II, book XXI (complete except for ch. 1); from the *Histoire de deux voyages* (fols. 29r–59r, 100r–102v, 113v–116r, 119r–122v); from the *Grand Insulaire* (I, fol. 260r–262r, 274v–275r).

1973

18 *Les Vrais Pourtraits et Vies des Hommes illustres (1584).* Facsimile ed. with intro. pp. (v–xiv) by Rouben C. Cholakian. New York: Delmar, 1973. 2 vols in–8°. A facsimile of the copy in the Widener Library, Harvard University.

1974

19 'Les Jumeaux', pp. 90–5 of Pierre Clastres, *Le Grand Parler. Mythes et chants sacrés des Indiens Guarani.* Paris: Éditions du Seuil, 1974. A reissue (after Suzanne Lussagnet, *Le Brésil*) of the *Cosmographie universelle* (II, fols. 919–20).

1978

20 *As Singularidades da França Antártica.* trans. Eugenio Amado. São Paulo: Livraria Itatiaia, 1978. 271 pp. in–4°.

1982

21 *Les Singularitez de la France Antarctique*, facsimile of the 1557/8 edition, with preface by Pierre Gasnault, in Jean Baudry, 'Un dossier Thevet'. Paris: Le Temps, 1982. 80 pp. + viii + 166 fols. + contents (2 fols.).

22 *Les Singularitez de la France Antarctique*, a Japanese trans. with notes by Ken-ichi Yamamoto, in *La France et l'Amérique*, (I). Tokyo: Librairie Iwanami Shoten (La Grande Découverte), 1982, pp. 157–501.

23 *Un'inedito di André Thevet*, critical edition of the 'Description de plusieurs isles', presented by Valeria De Longhi as a dissertation in the Università degli Studi di Milano (Facoltà di Lettere e Filosofia), Milan, 1981–2. Typewritten vol. of 434 pp.

1983

24 *Les Singularitez de la France Antarctique. Le Brésil des cannibales au XVIe siècle.* Selected texts, with intro. and notes by Frank Lestringant. Paris: La Découverte, 1982. 177 pp.

1984

25 *Voyages en Égypte (1549–1552). Jean Chesneau. André Thevet*, ed. and annotated by Frank Lestringant. Cairo: Institut Français d'Archéologie Orientale (Les Voyageurs occidentaux en Égypte), 1984. iv + 311 pp.

1985

26 *Cosmographie de Levant (1556)*, critical edition by Frank Lestringant. Geneva: Droz (Travaux d'humanisme et Renaissance, no. CCIII), 1985. cxxii + 374 pp. A facsimile of the 1556 Lyon edition (copy in Bibliothèque Mazarine: Rés. 16176) with intro., biblio., variations and notes.

1986

27 *André Thevet's North America. A Sixteenth-Century View*, edition in translation, with notes and intro., by Roger Schlesinger and Arthur P. Stabler. Kingston and Montreal: McGill and Queen's University Press, 1986. lx + 292 pp. Extracts from the *Singularitez de la France Antarctique* (chs 73, 74, 75–82); from the *Cosmographie universelle* (Books XXII and XXIII); from the *Grand Insulaire* (fols. 143r–159v; 176r–177v, 180v–183v and 403 r–407v); from the *Description de plusieurs Isles* (fol. 126r). Only the Canadian section of the *Grand Insulaire* (fols. 143–59 and 403–7) is given in French.

1993

28 *Portraits from the Age of Exploration: Selection from André Thevet's Les vrais pourtraits et vies des hommes illustres*, ed. by Roger Schlesinger, trans. by Edward Benson (Urbana: University of Illinois Press. Extracts from *Les Vrais Pourtraits*, ch. 52, 55, 66, 100, 101, 102, 141, 142, 145, 147, 149, 150).

Index

Abel, Cain and 65–6, 68
Adam 65
Aegean archipelago 121
aesthetics, Thevet's 32–5, 40
Africa 4, 8, 133
ageing, universal 98
Agricola, Rodolphus 60
agriculture, origins of 63, 68
Albaigne, André d' 177.n93
Albertus Magnus 30
alchemy 18
Alexander of Aphrodisias 62
Alfonse, Jean, of Saintonge 104,
    106, 107, 112, 131
  Thevet's use 122, 123, 126
allegory 77
Alvares, Manuel 126
Amazon River 54
Amazons 33, 71, 78–90, 113, 122,
    Pl.7
Ambrose, St 46
Americas
  interior of North 8
  named as Indies 30, 54
  see also Brazil; Canada, Patagonia
Anacharsis 105
anatomical texts 78
ancient authors
  compilations see Coelius
    Rhodiginus; Polydore Vergil
  philosophers 28–30
  Thevet's use of 12, 37–52, 62,
    63–4, 71–8
  see also individual authors
Aneau, Barthélemy 43–4
Angoulême; Thevet's birth in 9
Angoumoisine Mount 27, 31
animals, exotic 33, 34, 44, 46, 51,
    54, 76
  in Canon Lambert's map 113
  as emblems 42–3
  Thevet relates to Brazil 54

Annius of Viterbo 60
anonymous works
  Histoire naturelle des Indes 77
  Nouvelles des regions de la
    lune 19–20
  Satyre Ménippée 19, 20
Antarctic France 9, 53, 120–1
  see also Brazil
anthropology, discipline of 58, 69
Antilles 80
Antioch 45, 49–50
antipodes, Lactantius on 25
Antwerp school of cartography 105,
    119
Anville, Jean-Baptiste Bourguignon d'
    106
Apian, Peter 5, 104, 137.n19, Pl.2
Apinayé mythology 80
Arabia, Arabs 46, 55, 57, 87
Arawak tribe 80
architecture, origins of 63
Aristotle
  commentaries on 48
  elemental theory 21
  on sources of Nile 31
  on spheres of earth and water 8
  on torrid zone 30
  Thevet on 28, 29, 144.n76
arrangement of material
  reflects variety of subject-matter
    11, 29–30, 32–5, 125, 129–30
  transitions between topics 41–2
artefacts, native vii, 9, 76–7, 94–5,
    126
Asplenon 51
assay 18
assistants, Thevet's 39, 62–3, 67, 68,
    127, 128
astrology 6
atelier, Thevet's vii, 128
Athens 42
Athos, Mount 33, 44, 150.n64

Atkinson, Geoffroy 1, 2
Aubigné, Agrippa d' 20–1
Augustine of Hippo, St 4, 25
Aulus Gellius 44
autarky 99
authority
  of ancients 12
  and experience 11, 18–19, 24, 25
  questioned by dialogisms 49
autochthony 99
autopsy 13, 30, 59, 130
  see also experience
Avicenna 30
axe, Gê anchor-shaped 77
Aztecs 11

banana-palm 33
barbarians *see* savages
Barbaro, Ermolao 40, 49
Barberini, Cardinal Francesco 106
Bartholomew massacre 21
bears 33, 47
Belleforest, François de
  assistance to Thevet 39, 68
  attack on Münster 49
  criticism of Thevet 5, 12, 24, 28,
    54
  emblems 44
  on *kosmos* 144.n71
  on Marguerite de Roberval 122
  Thevet criticises 129
  varietas 130
Belle-Isle strait 117
Belon, Pierre 50
Bernard, Claude 18
Bethlehem 25, 42
Bible
  Creation story 7, 29, 65–6
  Luther's 50
  Thevet contradicts 25
  Wisdom 15
Bisselin, Olivier 126
blasphemy, cosmographers accused of
  7, 12, 24, 25, 130
Boaistuau, Pierre, of Nantes 38
Bodin, Jean 144.n73
Bordone, Benedetto 106, 121
Bourbon, Charles de, Cardinal of
  Bourbon 22
Bourdin, Gilles 22
bradype 54

Brazil
  Cabral's discovery 14
  fiction of New France 5, 7, 8, 9,
    53, 120–1, 124
  language 59–60
  Le Testu on 134–5
  monarchy 90–103
  mythology 37, 53–70, 71–103, 130
  religion 55–7
  'Thevet Island' 31, 119–20
  Thevet's journey to 9; fiction of
    earlier visit 27, 31
  Thevet's objects from 11, 126
  Thevet's use as paradigm 54–5,
    57–8, 58–60, 63–7
  see also Thevet (*Singularitez*);
    Tupinamba Indians; warfare
Breuning, Hans Jacob 50
bricolage, cartographic 111–13
Bruno, Giordano 25
Bry, Théodore de 16, 76–7, 95, 130,
  Pl.3
Buondelmonti, Christopher 109–10,
  121
Buxdorf, Johannes 149.n59

Cabbala 3, 39, 40, 41
cabinets of curiosities vii, 76–7
Cabot, John 119
Cabral, Pedro Alvarez 14
Cadmus 65
Cain and Abel 65–6, 68
Calepino, Ambrogio 38, 39, 81
California 116
Caloyer, reefs of 121
Camerarius, Johannis 48
Camerarius, Ludwig 5, 24, 28
Camers, J. 40
Canada
  England and 118
  fiction of New France 7, 8
  Roberval on 10
  Thevet as expert 120, 126
  'Thevet Island' 31, 120
cannibalism 55, 69, 76, 91, 94, 98,
  Pl.5
Cannibals, Promontory of 53
Cape Breton Island 119
Capella, Martianus 7
caraïbes (Brazilian prophets) 60, 63,
  66, 69

Cardano, Girolamo 3, 99, 100, 144.n77
Carib mythology 80
Cartier, Jacques 112
  absorption of discoveries into maps 113
  and Belle-Isle strait 117
  bilingual *colloquia* 59
  Le Testu on 135
  search for northwest passage 118
  Thevet uses nomenclature 119
cartography 37, 104–25
  bricolage 111–13
  fiction 108–16, 116–21, 124
  and geography 105–6
  integration of new discoveries 7–8, 108–9, 111–12, 115–16
  monsters 4, 113–14
  projections 108–9, 111–12
  use-value 105–6, 119
  see also portolans; rutters; *and individual cartographers and schools*
cartouches 113
Caspian Sea 57
Castiglione, Tommaso Porcacchi da 106, 109, 121
Catherine de Médicis, Queen of France 10, 22
Catholicism
  Council of Trent 24
  Counter-Reformation 56
  Gregor Horst on 50
  Index 154–5.n42
  Scandinavia 56
  Thevet and 5, 6, 10, 25, 56, 122–3
  see also blasphemy
Catullus 40
chameleon 53–4
Chaneau, Jean 47
Charaïbes (Brazilian prophets) 60, 63, 66, 69
Charles V, Holy Roman Emperor 14
Chateaubriand, François René, vicomte de 1
Chauveton, Urbain 4, 101
Chemlin, Caspar 49
China 34
Chinard, Gilbert 1, 17

Chios 39, 50, 51, 148.n51
chorography 2, 3–4, 5
Christianity
  juxtaposed with Brazilian religion 64–5, 66, 68, 71
  on man's possession of earth 29
  see also blasphemy; Catholicism; Greek Orthodox Church; meditations; Protestantism
Chronique de Nuremberg 113
Cicero 20, 40, 144.n76
classification 34, 35, 58
Clastres, Pierre 96–7, 130
clubs, Brazilian 73, 77
Codex Mendoza 5, 11, 126, 130
Coelius Rhodiginus, Lodovicus 37–41, 42, 46, 81
  and dialogisms 46, 47, 48
  emblematic borrowings from 44
  Horst notes Thevet's use 50–1, 61
  indexes 39
Coignet, Michel 107, 123
Coligny, Gaspard de, Admiral 10, 21, 93
  Le Testu and 104, 114, 116, 119, 132–3
Colines, Simon de 108
collection of exotic objects, Thevet's 11, 126
Colombia 80
colonization 16, 23–4, 130
  cartographical and literary fiction 5, 6–7, 8, 116–21, 122–3, 124
  English 6, 7, 8, 62
  French schemes 8, 9, 53, 93, 114, 116, 119, 120–1; failure 6–7, 10, 14, 16, 103, 124; fictions 5, 7, 8, 9, 53, 120–1, 124
Colossi, Colossians 33, 34, 42, 46–7
Columbus, Christopher 1, 13, 54, 80
commentaries 48, 50, 51
commerce, birth of 63
Constantinople 9
contamination of mythology 72, 82, 91, 99
Contant, Paul 76, 161.n20
cooking 69, 81
Copernicus, Nicolas 25
cosmetics and cosmography 32–5
cosmogony 7, 29, 63–7, 68–9, 130

cosmographical model 12–36
cosmography *see individual aspects throughout index*
Counter-Reformation 56
Creation, accounts of 7, 29, 63–7, 68–9, 130
Crete 33, 51, 65
crisis of cosmography, late Renaissance 129–30
criticism of Thevet 4–5, 10, 121, 129
 by Belleforest 5, 12, 24, 28, 54
 for blasphemy 7, 24, 25
 for cartographical fictions 108
 over Indian monarchy 93, 96, 101–3
Cryb tribe 156.n51
Cuba 109
Curco, island of 121

Daedalus 65
danger, personal 13–14
Dares Phrygius 62
Dead Sea 31–2
Delaune, Étienne 73–6, 77, 78, Pl.4
Deleuze, Gilles 68, 104
deluge myth 66, 68, 69, 98
'Demons, Isle of' 122–3
dialogisms 46–8, 49
Diderot, Denis 102
Dieppe school of cartography 113, 114, 116, 119, 124
diet 57
 raw/cooked distinction 69, 81
Diodorus Siculus 68
Diogenes Laertius 142.n54
Dioscorides 31
discoveries, new
 accommodation within existing cosmographical framework 7–8, 108–9, 111–12, 115–16
 fictions 116–21
distance from subject 30–1
Dorat, Jean 9–10, 22, 23
doubling-up, fictive 31
Drake, Sir Francis 177.n93
dress, Brazilian 73, 76, 73–4
 regalia 94–5
du Bartas, Guillaume de Salluste 72, 161.n20
du Bellay, Joachim 9–10
du Moulin, Antoine 18

duplication, fictive 31
du Préau, Gabriel 25
Duverger, Christian 130
dynamism of cosmography, morphological 3
 accommodation of new discoveries 7–8, 108–9, 111–12, 115–16
 see also fiction

ecstasy 19, 20–2
 see also euphoria
Egypt 31, 34, 40, 42, 44, 45, 49, 65
elders, Tupinamba 101–2
elements, four 21
elephants 46, 51
emblematics 41–5, 72, 78
Empedocles 28, 29
England
 colonization 6, 7, 8, 62, 118
 and continuous ocean theory 111
engravers, Thevet's vii, 119
entropy 100
equator, crossing of 16–17
Erasmus, Desiderius 44–5, 46, 50
Erondelle, Pierre E. 167.n84
erudition, display of 37–41, 47, 50, 51–2
Estienne, Robert and Henri 39, 81
Euhemerism 62, 64
euphoria
 over cosmographer's view of world 15–16, 16–17
 of navigator at sea 15, 26, 62
Eusebius 65
eusunopton 115
exaggeration 31–2
ex-libris, Thevet's 126
'exoticism' 1
experience 12, 13, 15–16, 17–19, 128
 and authority 11, 18–19, 24, 25
 hypothetical 33
 and theory 17–18, 26, 104, 111, 120, 123–4, 131
Ezekiel 7, 50

Falkland Islands 122, 130, Pl.10
Fathers of Church 25
fauna
 of Brazil 9
 see also individual creatures and animals, exotic

fiction 8–9, 33
  cartographic 108–16, 116–21, 124
  of colonial possessions 5, 6–7, 8, 116–21, 122–3, 124
  duplication 27, 31
  of erudition 37–41, 47, 51–2
  exaggeration 31–2
Finaeus, Orontius (Oronce Finé) 6
Fioravanti, Leonardo 18–19
fire, origin of 68–9
fish, respiration of 47
Flaubert, Gustave 17, 129
Fleece, Golden 23
Flemish school of cartography 105, 119
Flood myth 66, 68, 69, 98
Florida 8, 10, 126
France
  Antarctic 5, 8, 9, 53, 120–1, 124
  and Southern Land 8, 118
  Wars of Religion 24
  see also under colonization
Franciscan Order 9, 10
Friederici, Georg 80, 165
Frobenius, Leo 37
Frobisher, Martin 111
fruits, exotic 76

Gaguin, Robert 60
Galileo Galilei 24
Gandía, Enrique de 165
Ganges River 54
Garcie, Pierre 107, 123, 126
Gaza 45–6
Gê people 69, 77
Génébrard, Gilbert 5, 121
Genesis, Book of 7, 29, 65–6
geocentrism 3, 21, 25
geography, new discipline of 105–6, 129
Gérando, Joseph-Marie de 58
gigantism 94, 98–100, 122, 135
Gilles, Nicole 128
Gilles, Pierre 46
Giovio, Paolo 59
giraffe 42–3
glacial zone 8, 29
Gohory, Jacques 18
gorgons 79
Grail, holy 55, 68
Greek Orthodox Church 50

Guanabara 53
Guattari, Félix 68, 104
Guayaki of Paraguay 96–7
Guiana 80
Guise family 10
gulfs, catalogue of 54

habitability of earth 8, 29, 30
Hacket, Thomas 61–2, 143.n57
hair, shaving of 63
Hakluyt, Richard (the Elder) 6
Hakluyt, Richard (the Younger) 6, 7, 24, 111, 130
hare, etymology of 33, 40, 42, 47–8
head, shaving of 63
Henri III, king of France 10, 14
'Henryville' (Ville-Henry) 5, 121, 124
Hercules 71
heresy 6
Héret, Mathurin 62–3, 67, 68
Herodotus 31, 47, 79, 84
Hippocrates 144.n76
hippopotamus 54
Histoire naturelle des Indes (anon.) 77
histories 33–4, 64, 105, 106
homogeneity of world, increasing 100
Hondius, Jodocus 6, 45
honey 40, 80–1
'horizons, new' 1, 2
Horst, Gregor 37, 48–52, 61, 148.n47, 148–9.n54
hubris, cosmographers' 5–6, 10, 15, 28, 130
humanism 18
Humboldt, Alexander von 163

Ibn Sina (Avicenna) 30
illustrations
  iconography 5, 103
  reuse in inappropriate context 87–9
  see also individual artists and under individual works
imagines mundi, medieval 5
Imaugle (Imangla) and Inébile, islands of 84–5, 86–7, 88, 122
incompletion of Thevet's works vii, 128

Indagine, Jean d' 18
Index, Roman Catholic 154–5.n42
index to works of Coelius
    Rhodiginus 39
Indies, Americas named as 30, 54
Inébile *see* Imaugle
instruments, navigational 16, 19
insulist, Thevet as 26–7, 110–11,
    Pl.1
inventors, mythological 64–5, 66
inversion 79
Isidore of Seville 4, 113
Islam 55–6, 57
islands 106–8, 109–12
    amazons associated with 81–2,
        84–5, 86–7, 88, 122
    fragmentation of cosmos into 107,
        121–5
    'insulist' 26–7, 110–11, Pl.1
    singular nature 35
    'Thevet Islands' 31, 119–20
    see also individual islands and
        Thevet, André (*Grand Insulaire*)
isolario genre 106, 109–10, 124

Jacquard, Antoine 76–7, 78, Pl.5
Jarativa 80
Jason 10, 23
Java-la-Grande 115, 116, 124, 130, 134
Jericho 42, 45
Jerome, St 47, 50
Jerusalem 9, 25
Jesus, Society of 49–50
Jodelle, Étienne 9–10, 22
Johanna, Mother (superior of
    Hospital of Santi Giovanni e
    Paolo, Venice) 85
Jonah and the whale 7, 25
Josephus, Flavius 65
Judaea 42
judge, Egyptian 44

Kabbala 3, 39, 40, 41
Kircher, Rudolph 50
Konyan Bebe 91
kosmos (Gk.) 32

la Boderie, Guy Le Fèvre de 22, 23
la Boétie, Étienne de 166.n83
labour, Tupinamba division of 83, 84
labrets (lip-ornaments) 73

Lactantius 25
Lambert, Canon of Saint-Omer 113
languages
    foreign, in works on cosmography
        59–60
    Latinisms 94
    vigour of Thevet's 128
la Nuche, Louis Guyon de 151.n5
la Popelinière, Lancelot Voisin de 4,
    8, 24, 101, 118, 124
la Porte, Maurice de 141.n29
la Rochefoucauld, François III, comte
    de; epistle to 32, 40, 104
las Casas, Bartolomé de 38
Latinisms 94
latitude 111–12, 115
la Tourette, Alexandre de 18
Laudonnière, René de 10
Lemaire, Jean, of Bellegem 60–1
Le Roy, Jean, sieur de la Boissière
    76
Léry, Jean de
    on amazons 89–90
    bilingual *colloquia* 59
    on de Bry 77
    iconography 103
    on Indian monarchy 90, 91, 93,
        96, 97, 101–3
    on origin of fire 69
    polemic against Thevet 24, 121
    on savages 57, 58, 67, 95
    sensual writings 17
Lescarbot, Marc 94, 95, 121, 131
Le Testu, Guillaume 104–5, 108,
    116–18, 132–5
    on Africa 133
    on Americas 121, 134–5
    articulation of elements of map
        108, 115–16
    and Coligny 104, 114, 116, 119,
        132–3
    fictious lands 116–18
    monsters 4, 113–14
    on Southern Land 115, 116, 124,
        133–4
    use-value of maps 119
Leu, Thomas de 110–11, Pl.1
Levant 9, 59
    see also individual places and
        Thevet, André (*Cosmographie de
        Levant*)

Lévi-Strauss, Claude 53
  on amazons 80, 165
  on cooking 69
  on global entropy 100
  iconography 103
  and Indian cosmogony 130
  on women in Brazil 83
l'Hospital, Michel de 22
Lindsay, Adam 106, 107
lips, perforated 72, 73, 74
Lithuanians 56
Loaysa, Garcia Jofre de 176.n85
longitude 54, 111–12
Lorraine, Cardinal of (Charles Guise) 22
Louis IX, king of France 94, 98
Lucian 20
Luther, Martin 50, 59
Lutheranism 5, 56
lycanthropy 33, 47, 149.n54

Macrobius 20
Maffei, Raphaello (Volaterranus) 47
Magellan, Ferdinand 98, 100, 113, 122, 135
  Staden's engraving 16, Pl.3
Magellan Strait 117, 134
Maino *or* Mayno, Jason de 60
Mairs (Tupinamba civilizing heroes) 56, 60, 61, 64, 66, 67
Malvinas 122, 130, Pl.10
man of nature 1, 58
Mantua 39, 42
mappae mundi 2, 3, 105, 106
Maranhâo 76
Marco Polo 112–13, 163
Margageats 59, 72, 75–6, 88, Pl.6
marginalia 49, 107–8, 127–8
Marguerite de Navarre 122
marriage, laws of 63
Marseille 33, 40
mathematics 6
Mattioli, Lodovico 31
Mauro, Fra 163
Medina, Pedro de 15
meditations, cosmographical 6, 45–6
memory, art of 3, 62–3
Ménard, Pierre 127
Menippean satire 19, 20
Mercator, Gerardus
  cosmographical meditations 6, 45

innovations 111, 119
religious purpose 6, 7
world-map (1569) 111, 175.n63
Merlin the Sorcerer 55
method, Thevet's working 127–8
Meung, Jean de 94, 95
Mexico 11, 130
milestones (*padröes*) of Portuguese navigators 4
Mizauld, Antoine 6
Moluccas 15, 109
monarchy, Indian 64, 90–103
monsters 4, 113–14
Montaigne, Michel Eyquem de
  'On Cannibals' 57, 58, 94
  and cosmographical model 12
  'On Experience' 18
  on imagination 114
  and Noble Savage 1
  on political development of New World 97, 98
  sources of quotations from ancients 38
  structure of *Essais* 127–8
  on topography 26
  on variety of world 29, 32
monuments, ancient 34
moral, physical as reflection of 94
morbicha (Brazilian; 'elder') 59–60, 102
mummy, Egyptian 34, 49
Münster, Sebastian
  Belleforest attacks 49
  commentaries on ancient authors 12, 40, 48
  and Ptolemy 12
  religious purpose 6, 7, 45
  renovates cosmography 104, 129
  Thevet's use 126, 128, 153.n24
mythology 71–103
  on amazons 78–90
  ancient authors 71–8
  Arabian 87
  on Brazil 37, 53–70, 71–103, 130
  comparative, Thevet and 55
  contamination between cultures 72, 82, 91, 99
  Creation 7, 29, 63–7, 68–9, 130
  Flood 66, 68, 69, 98
  Indian monarchy 90–103
  on inventors 64–5, 66

mythology (*continued*)
  local, in chorographies 4
  monsters 4, 113–14
  and politics 90, 91
  and warriors 71–8

Nacol-Absou, king of the Promontory of Cannibals 53, 64
Namikwara family 103
'natural questions' 33–4
nature
  Le Testu's eulogy 105
  man of 1, 58
navigation
  Brazilian and Caspian 57
  eulogy of 15, 26, 62
  Thevet's use of technical writings 11, 123
  see also experience; portolans; rutters
Newfoundland 111–12, 117, 135, Pl.9
Nicolay, Nicolas de 50, 106, 123
Nile, sources of 31
north-west passage 8, 116, 118
Nostradamus 121
Nouvelles des regions de la lune (anon., 1604) 19–20
Nuremberg Chronicle 113

objects *see* artefacts
ocean 3, 14–16, 28, 56
  enclosed or encircling 8, 27–8, 30, 111, 124
Olaus Magnus 55, 56
onomastic play 34–5
Orellana, Francisco de 82
Orpheus 23
Ortelius (Abraham Ortel) 105, 119, 112, 175.n63
ostrich 31
Other, the 1, 92, 96
Ottoman Empire 1–2, 50
Ovid 65
Oviedo, Gonzalo Fernandez de 12–13, 14

padröes 4
paganism 55–7, 63, 65
Pagès (healers of Tupinamba) 55, 63, 66

Palissy, Bernard 12, 18
Panofsky, Erwin 71
Paracelsus 18
Paracoussi, king of La Plata 64
Paradise, Islamic 55
paraenesis 44, 48
Paraousti Satouriona 64
Paré, Ambroise 18, 145.n9
Pascal, Blaise 30
Patagonia 98–100, 122, 126, 135
Peiresc, Nicolas-Claude Fabri de 108
Perotti, Niccol 38, 39, 81
Peter Martyr d'Anghiera 63, 80
philology 37–52, 42
philosophy
  ancient 28–30
  natural 5, 32
Phoenicians 65
Phrygia 65
physical as reflection of moral 94
Pico della Mirandola 39
Piero di Cosimo 76
Pigafetta, Antonio 82, 98, 106, 107–8, 123, 168.n101
pilots *see* navigation
plagiarization 127, 128–9
Plantin, Christopher 152.n15
Pléiade (poetic group) 22
Pliny
  Barbaro's *castigationes* 49
  on Colossians 47
  commentaries on 40, 47, 48
  on inventors 65
  on *kosmos* 144.n71
  on monsters 4
  ostrich 31
  Oviedo's use 12–13
  popularity of works imitating 48
  on prodigies 1
  Rabelais' use 127
  Solinus' use 51
  on sources of Nile 31
  Thevet's use 33, 40, 47, 62
Plutarch 11, 39, 157.n56
poetry 21, 22–3, 23–4, 34
point of view, cosmographer's 5, 28–9, 30–1
polemic against Thevet *see* criticism
politics
  and cartography 116–21

Indian monarchy and 90–3, 97
Wars of Religion 24
see also colonization
Poliziano, Angelo 43
Polo, Marco 112–13, 163
Polydore Vergil *see* Vergil, Polydore
Pomponius Mela 1, 12, 40, 49, 152.n24
portolans 8, 105
Portugal
cartography 113, 114
cosmographer royal 10
explorers 4, 122
Southern Land hypothesis 116, 117, 124
and Treaty of Tordesillas 3, 14–15, 116
war of succession 8, 118
Postel, Guillaume 6, 85
projections 108–9, 111–12
Protestantism 7, 10, 24, 50, 59
pseudomorphosis 71, 78
Ptolemy (Claudius Ptolemaeus) 3, 4
geocentricism 3, 21, 25
Münster's edition 12
Thevet's use 7, 12
and Thule 114–15
Purchas, Samuel 130
Purgatory 56
Puritans 7
Pygmies 7, 41, 50, 113
pyramids 42

quality/quantity 2–3, 5, 30
Quesada, Fernan Pérez de 80
Quoniambec (Tamoio chief) 53, 64, 91–103, 166.n73, Pl.8

Rabelais, François 37, 38, 122, 127, 141.n29, 152.n15
racism 58
Ramusio, Giovanni Battista 2, 130
Rats, Isle of 26
raw/cooked distinction 69, 81
Razilly, François de 76
reading process 49
regalia, Quoniambec's 94–5
regeneration of world 98
regression 81
religion
Aztec 11

Brazilian 11, 55–7, 64–5, 66, 68, 71
cosmographical meditations 6, 45–6
'Isle of Demons' fiction 122–3
mystery of 28
see also Christianity; mythology; theology
reversals, hyperbolic 79
reworking of material 87–9, 126, 128
rhizomatic relations 68
Ricchieri, Lodovico *see* Coelius Rhodiginus
Richer, Pierre 121
Rio de Janeiro 53, 120–1
risk, personal 13–14
Roberval, Jean-François de la Roque de 10, 118, 135
Roberval, Marguerite de 122
Ronsard, Pierre de 9–10, 21–2, 23
rutters 7–8, 105, 106

Saguenay 118, 135
Saint-Germain-des-Prés 98
Samson and the lion 7, 25
'Sanson, Isles of' 122, Pl.10
Sansovino, Francesco 2, 59
Saqqara 34
satire 19–20
Satyre Ménippée (anon.) 19, 20
Saumaize, Bénigne 142.n51
savages
Noble 1, 58
Thevet's affinity for 11
Thevet's use as paradigm 55, 57–8, 66–7
see also amazons; cannibalism; Tupinamba Indians; warfare
savant/navigator relationship 18, 104, 120, 131
saw, invention of 63
scale 2–6, 19
cartographic 108–9, 130
cosmography comprehends local and universal 5–6, 26, 30–1, 35
and distance from subject 30–1
transgression 21, 26–7, 27–8
Scandinavia 55, 56
Schedelsche Weltchronik 113
Schilder, Valentin 149.n59

scientific method 33–4, 49, 58
Scythia 79, 84, 105
sea *see* ocean
Servetus, Michael 25
sexes; transgression of normal roles
    84
    see also amazons
similarities, interethnic 55
Simler, Josiah 45
simplification
    in cartography 114–15
    through generalized exchange 100
Sinai desert 31
singularities
    collections of 32, 126
    impossibility of classification 35,
        58
    natural philosophers' interest in 5
    on savages 57–8
Société des Observateurs de
    l'Homme 58
society, origins of 63–7
Solinus 1, 12, 40, 41, 51
sources, Thevet's 10, 11, 26, 127,
    128–9
    personal collection 11, 126
    technical writings of navigators
        11, 123
    see also individual authors and
        ancient authors; artefacts
Southern Land 8, 110–11, 115, 124
    French interest in 8, 118
    Le Testu and 114, 117, 124,
        133–4
    Portuguese originate 116, 117,
        124
space, Brazil as paradigm of distant
    54
Spain
    cosmographer royal 10
    Treaty of Tordesillas 3, 14–15,
        116
Sparta 39, 40
sphere, cosmographer's celestial
    21–2
Staden, Hans 16, 77, 91, 97, Pl.3
Strabo 46, 47
Strongile, islet of 121, 164.n56
Strüppe, Joachim 49
symbolics; *Cosmographie universelle*
    22

Tabajares 72, 75–6, Pl.6
Tabourot, Étienne 38
Taprobana 109, 110
theology 5, 25
    see also Bible; blasphemy;
        Christianity; meditations
theory/practice relationship 17–18,
    26, 104, 111, 120, 123–4, 131
Thevet, André 9–11
    Cosmographie de Levant;
        dedicatory epistle 32, 40, 104;
        dialogisms 46–8; diversity of
        subjects 33; poetry quoted at
        opening 22; popularity 126–7;
        publication 9; sources 45,
        37–41, 46–8; Vadianus as
        model 127
    Cosmographie universelle; on
        amazons 84–5, 85–6, 87–9;
        ancient authors used in 48; on
        Brazil 53, 67, 72, 83, 91, 94; on
        comprehensibility of world to
        navigator at sea 15; dialogisms
        48; illustrations Pls.6,7,8; on
        mathematics 6; on personal
        dangers 13–14; plates 87–9,
        94; poetry quoted at opening
        22, 24; preface 14, 32–3; and
        Ptolemy 7, 12; reworking of
        material 128; sources 11, 48;
        symbolics 22
    Description de Plusieurs Isles 126,
        128
    Grand Insulaire et Pilotage
        106–8, 110–12; amazons 122;
        atomization of world into islands
        107, 121–5; cartographical
        framework 110; disorder
        106–7; fiction of new-found
        lands 118–21; illustrations
        Pls.1,9,10; magnetic lines of
        direction 111; maps on Brazil
        53; Newfoundland 111–12, Pl.9;
        reworking of material 126, 128;
        sources 10, 37, 107–8; use-value
        106; variety of world 125
    *Histoire des deux voyages aux
        Indes australes et occidentales*
        67; difficulty of reading 123;
        practice and theory juxtaposed
        123; Quoniambec 91, 96;

reworking of material 53, 126, 128; sources 10

Singularitez de la France Antarctique 9, 53–70; amazons 71, 81, 87; Brazilian warriors 72; Christian perspective 64–5, 66, 68, 70, 71; on error of ancients 29; fiction of Antarctic France 121; Héret's contribution 67; illustrations 9, 82, 83, 87–9, Pl.7; mythology 37, 53–70; poetry quoted 22; use of Polydore Vergil 61–70, 71, 127; Quoniambec 91, 166.n73; reworking of material 87–9, 128, Pl.7; variety and arrangement of material 33–5

Vrais Pourtraits et Vies des hommes illustres; comparison of old and new worlds 97; *Grand Insulaire* announced in 106; illustrations 5, 94, 95; Indian monarchs 53, 64, 72, 93–5; poetic quotation 22; popularity 127; pun on Lemaire 60–1; sources 11

see also individual topics throughout index

Thevet Islands 31, 119–20
Thou, Jacques-Auguste de 108
Thule 56, 114–15
Tiphys 23
tobacco, origin of 81, 82
topography 26, 81–2
Tordesillas, Treaty of 3, 14–15, 116
torrid zone 8, 29, 30
transgression
  of scale 21, 26–7, 27–8
  sexual 84
transition, cosmography in state of 129
transitions between topics 41–2
translatio imperii 98
Tree of Knowledge 33
Trent, Council of 24
Trophony 56
Tupinamba Indians 53
  cannibalism 69
  elders 101–2
  lips, perforated 72
  monarchy 90–103

as paradigm of barbarians 55
religion 11, 130
women 83, 84
Tupinikin people 82–3
Turks 1–2, 33, 44, 50, 55–6, 153.n27

unity of cosmography 104–8
universalism 11, 25, 28–9
utopia, Léry's social and military 101–3

Vadianus, Joachim 6, 40, 45–6, 127, 155.n42
variety 32–5, 40, 50, 100, 114, 130
  and classification 34, 35
  Thevet reflects 11, 29–30, 32–5, 125, 129–30
Vaudeclaye, Jacques, of Dieppe 121
Velho, Bartolomeu 177.n93
Venice 2, 85, 106
Vergil, Polydore 61–70, 71, 73, 74, 127, 154–5.n42
Versins, miracle of 57
Vesalius, Andreas 77, 78
Vespucci, Amerigo 113
Villegagnon, Nicolas Durand de, Knight of Malta 9, 62, 95, 101
Ville-Henry 5, 121, 124
Vinet, Élie 128
Virgil 23, 65
vision, eulogy of 43
Volaterranus (Raphaello Maffei) 47
volta, South Atlantic 14

Waldseemüller, Martin 104
warfare, primitive 63, 71–8, Pls.4,6
  war-chiefs 90, 95, 96
  women's role 83
Warrau mythology 80, 82
Wars of Religion 24
weapons, Brazilian 76, 77
winds, trade 14
Wisdom, Book of 15
women; role of primitive 83, 84, 85
  see also amazons
workshop, Thevet's vii, 128

Yérasimos, Stéphane 2
youth, fountains of 153.n24
Yucatán 80